Dad, Why Are You A Global Warming Denier?

A Short Story That's Right for the Times

by Bob Tisdale

INTRODUCTION

Note: This is the print edition of my short story *Dad, Why Are You A Global Warming Denier?* The body of the print edition and the Kindle editions are the same, with the following exceptions: The Amazon Kindle edition of this book contains hyperlinks, which will not work in this print edition, for obvious reasons. I've left the text underlined and in blue where the hyperlinks existed. In this print edition, the text associated with those hyperlinks has been numbered (superscripts), and the http addresses for those links are listed at the end of the book, under the heading of REFERENCES. If you want working hyperlinks, you'll have to purchase the Kindle edition. In a number of places, I've also had to shift where the illustrations landed in the text to minimize the number and sizes of blank space.

# # #

I penned the following short-story, set in the year 2017, as a series of conversations between a twentysomething daughter and her mid-50s father. The father is a heretic of the church of catastrophic human-induced global warming/climate change. The daughter, Anna, is a believer.

This work includes quotes from the IPCC (Intergovernmental Panel on Climate Change) and a well-respected politician. It also includes 22 illustrations, most of which are in color.

As you will see shortly, it is written in the first person by Anna.

###

DAD, WHY ARE YOU A GLOBAL WARMING DENIER?

As I write this, the year 2017 is drawing to a close. Weather around the globe is weird. The President of the United States is a global warming denier, and so is my father. Dad's an engineer by training, as am I, but a couple of years ago he changed vocations to real-estate agent. I believe it timely to relate the following recent exchanges I had with him about catastrophic human-induced global warming and climate change.

Because the topics are technical and political, my writing may seem a bit stilted at times. But I've reproduced what my father said as best I can and included the graphs he presented. He can come across as nerdy, because of his engineering training and his constant reading and studies for his own understandings. Where many other dads hunt and fish and golf and watch TV sports, my dad reads constantly and investigates when he can, especially when a topic interests him.

So, let us return to that stormy day when the conversations began.

# # #

I left my rain-soaked raincoat, my soggy floppy hat, and my inside-out umbrella on the hall tree in my parents' foyer. On the floor, I dropped my drenched and drippy backpack, which contained my now-soggy stuff for the weekend.

"Dad, I'm home," I hollered.

"Up here, honey," came Dad's bellowed reply, muted by a closed door and distance.

I marched up the stairs to his second-floor man cave, each step announcing my approach.

As anticipated, before I could knock, he called, "Come on in, Anna."

I found Dad standing before one of the windows, smiling as he looked out on the gale-driven rain lashing our yard. Turning and seeing my soggy state, he walked quickly over to me, his long legs covering the distance in two giant strides. Wrapping me in his arms and kissing my forehead, Dad said, "What a great surprise, Anna."

"Something's been bugging me and I needed to come home for a few days."

"Everything okay?"

"'Everything' as in life in general? Yeah. Everything's fine."

"Great! You missed your Mom, though. She's at Gram and Gramps' and should be home Sunday. At least, that's the plan."

"I know. I spoke to her yesterday. She said you'd be home alone. So I came to keep you company."

"Thank you so very much, sweetie. I'm showing a duplex in town tomorrow, and I couldn't get away. Now I discover it was a blessing."

"Your couldn't-get-away excuse usually means Uncle Marty would be at Gram and Gramps'."

"I had lunch with Gram and Gramps on Tuesday. I was in that neck of the woods for a meeting that morning. So we've visited recently. Speaking of lunch, have you eaten?"

"I grabbed some pizza before I got on the train. Thanks, though."

"A cup of tea or a hot chocolate?"

"Tea sounds great, Dad."

"How come you didn't call so I could pick you up at the station?" He gently wiped some residual rainwater from my face with his thumbs.

"I wanted it to be a surprise."

"And it's a great surprise. Come on. Let's sit in front of the fire. Warm you up some."

In response, I exclaimed, "In addition to being stormy, it's freakishly cold. Thank you, human-induced climate change."

Dad rolled his eyes in feigned disgust. "Oh, Anna, not that crap again."

"'Oh, Anna, not that crap again,' what?" I said with a smile. Our opinions on global warming and climate change were worlds apart. We normally skirted the subject, and one of the reasons my Dad avoids Uncle Marty is global warming is a hot topic with my mother's brother.

A bit of background: Dad used to be a mechanical engineer, but left it after more than two decades, to go full time into real estate. With me, his only child, almost out of college at the time, he wanted to take a shot at something he "always wanted to do". He was a natural and is now making more money than he ever did working at the engineering firm. Me, I'm following in his footsteps in engineering. The big surprise: three years ago, a few years after he retired, I was hired by the firm he had worked at for so many years, hired before I graduated.

Dad cupped my face in his hands and stared heavy-duty into my eyes. "You come across like one of the indoctrinated multitude who believe in that anthropogenic global-warming and climate-change nonsense. Just like your Uncle Marty."

"Nonsense? I have a question."

"Uh-oh," he said, then relaxed his eye contact and gently broke the embrace.

"Dad, why are you a global warming denier?"

"Ah, so that's what's bugging you. You think I'm a global warming denier."

"Yup."

"Anna, that global warming nonsense is nothing but global politics at its worst," he said, now standing at the man cave's kitchenette, putting on a pot of water to boil for tea.

Raising an eyebrow in disbelief, I grabbed a spot on the couch before the fireplace and asked, "And on what are you basing your opinion, o' opinionated one? With today's storm, almost everyone on the train was talking about climate change, and they were *not* praising it."

"The simpletons who use the word *climate* when discussing a single weather event do not understand the definition of *climate*. It's not difficult; according to the World Meteorological Organization, climate is the 30-year average of weather. And my statement about the global-warming and climate-change nonsense being global politics at its worst is not opinion. There's evidence in plain sight, sweetie."

In an overacted display, I scanned the room and cleared my throat.

Unimpressed with my performance, he continued, "You only have to look at the lawsuits brought against oil companies, the ones filed by state attorneys general and by shoreline cities in California. Those lawsuits should lead anyone to conclude that one of the many planned offshoots of the global warming scare was to allow the politicians and their lawsuit-filing henchmen to continue their wars on fossil fuel concerns that've had strangleholds on the world's economies for decades."

"My apologies for my earlier name calling, Dad, but rising sea levels *are* threatening the cities of Oakland and San Francisco *and* all low-lying coastal properties around the globe."

Dad placed our teas on the coffee table and sat down beside me on the couch before the hearth. "Are oil companies' liable for unknown future damages because city founders located their cities well below the paleoclimatological high-tide mark, Anna? Of course not. City founders located the bayside cities of Oakland and San Francisco next to San Francisco Bay for economic reasons. The Pacific Ocean was and continues to be used for maritime trade, the lifeblood of those cities."

It always amazed me how well-spoken and how conversant in technical and scientific terms he could be. It meant he had studied a topic and studied it well. I've always tried to emulate those traits in my father. Understanding topics and speaking well helped me in high school and college, and I believe they helped me land the job I have.

And thoroughly my father's daughter, I was usually right there with him, equally informed about these topics, or so I thought. But I had to ask, "Rewind a moment, and please explain 'paleoclimatological high-tide mark'."

He mimicked talking backwards, making me giggle. He then put back on his serious face. "Happy to explain, Anna. I think I just coined it so don't bother to look online. Scientists who study ancient climates, called paleoclimatologists, use proxies of all sorts to estimate what climate was like in past eras, sometimes to hundreds of thousands and millions of years ago. Geologists have also performed paleo-sea-level studies. Decades ago, it was determined that, between ice ages, during periods known as interglacials, sea levels rose to levels that were 20 to 30 feet higher than they are today. And in some estimates of the Pliocene, about 3 million years ago, they were 40 *meters* higher."

I had read about rising sea levels in numerous news articles, so I said, "And that 20 to 30 feet sounds like the estimated range of how high sea levels will rise in response to our carbon dioxide emissions."

"It's no coincidence, Anna. There is only so much water stored on land in glaciers and ice sheets and only so high sea levels can rise when and where and *if* it melts. I say *if* because sometime in the future we'll return to another ice age. When another ice age begins, snow and ice will accumulate on land surfaces, and sea levels will drop again. Back to now: regardless of whether or not there is even a measurable human impact on sea levels, the oceans are going to rise and consume everything in their way until Earth's climate starts to cool back towards another ice age. More to the point, global temperatures have been above the melting point of glaciers and ice sheets ever since the end of the last ice age—about twenty thousand years ago—and sea levels have risen three to four hundred feet in those twenty thousand years. Cutting greenhouse gas emissions is not going to stop sea levels from rising. No way, no how. Electric cars, wind turbines, and solar cells can only aid against rising oceans if we were to crush them and use their dead bodies as landfill for levees and dykes."

Much as I hated to admit it, I was agreeing in part with him. "That long-term look makes sense, Dad."

"Of course. That's the paleo-data-based reality, like it or not."

"I keep forgetting that sea levels were 300 to 400 feet lower at the peak of the last ice age."

Dad said, "Not to put words in our mouth, but...you appear to also keep forgetting that sea levels have risen that 300 to 400 feet since then because global temperatures have been above the melting point of glaciers and ice sheets for 20,000 years. *And* that sea levels will continue to rise another 20 to 30 feet regardless."

Feeling somewhat deflated, I interjected some levity, "But Uncle Marty will not be pleased when I tell him you want to send his EV to the car crusher."

"It's the car crusher now or after it burns itself to the ground."

With news stories appearing in my mind's eye, I said, "But, Dad, the sea levels are only rising at the present accelerated rate because of our CO_2 emissions."

"Really, Anna? Are you sure? Based on satellite altimetry data, the current rate of sea level rise is roughly one foot per hundred years, maybe a little more. Gloom and doom mongers regularly compare two types of sea level data—long-term, tide gauge-based data; and short-term, satellite-based data—and claim that the rise in global sea level is accelerating."

"Whoa there, Dad. 'Satellite altimetry data', paleo high-tide levels. What's with the high-tech speak?" I knew the answer, but I had to ask anyway.

"I'm not trying to impress you, Anna. I've been studying this for decades, and the lingo's stuck in my head."

But Dad surprised me when he suddenly rose, then grabbed his tablet computer from the nearby oak desk and returned to the comfort of the couch. His fingers danced on the keyboard.

After a moment's wait, he said, as he handed the computer to me, "I was focusing on sea levels a few years ago. This graph compares the trends of global tide gauge-based data from 1855 to 2001 *with* global satellite-based sea level data, which, at that time, only ran from 1993 to 2013."

"Looks familiar. I've seen graphs like this before. In fact, I've seen them many times."

"Alarmists publish graphs like this all the time, because the trends are quite different. The satellite-era data have a trend of 3.2 mm/year, which converts to a little more than one-eighth of an *inch* per year, while the long-term tide-gauge data have a trend of about 1.7 mm/year, slightly more than one-sixteenth of an inch per year. That is, the trend of the long-term data is about half that of the satellite-based data."

I threw a verbal jab. "Looks like the rise in sea levels is accelerating, doesn't it?"

"Yeah, yeah, yeah. But only to the casual observer. There are lots of ways those comparisons can be misleading."

Below is Dad's first sea level comparison graph, presented as my Exhibit 01.

Exhibit 01

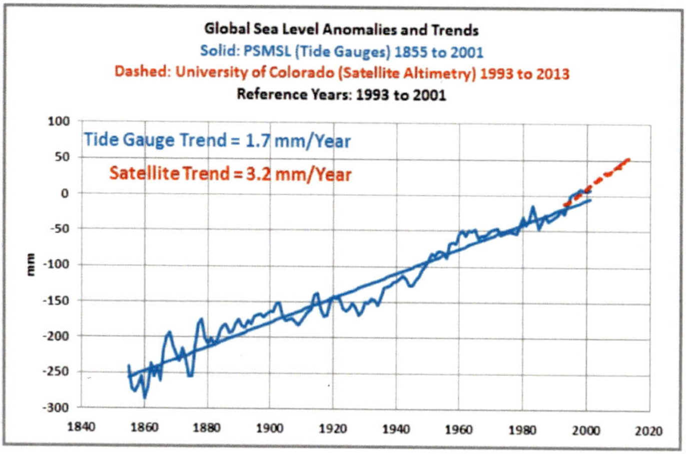

"If you want, later on, I'd be happy to discuss why graphs like that can be misleading, but there's another point I want to make right now. Take a closer look at the long-term tide-gauge data. There's something very obvious in that curve. The rise in sea level in the long-term tide gauge data has not been steady. See how the data modulate back and forth across the linear trend line? There are multidecadal periods when the sea level rise accelerates and other multidecadal periods when the rise decelerates, even falls. With the trend values and trend lines shown on the graph, people tend to focus on them and not the long-term wiggles."

Looking again at the graph Dad furnished, it was as he described. The tidal gauge-based data do show periods when the rate of rise accelerated and others when it decelerated.

Dad paused to collect his thoughts. Then he explained, "When I put together that graph a couple of years ago, they only had twenty-one years of the satellite-based sea level data. That data are available starting in 1993."

"'Data are' has always sounded weird to me, but I know the grammar's correct because the word data is plural of datum."

"Except when you're talking about Data from *Star Trek: The Next Generation.* He's singular."

"Not bad, Dad. On a scale of one to ten for Dad jokes, that was a seven."

"Okay, back to sea level data: We know the trend for that dataset from 1993 to 2013 was about 3.2 mm/year. However, because of the multidecadal variations in the tide gauge-based data, which began in 1855 and ended in 2001, I wanted to determine the 21-year trends for that dataset as well. That way

I was comparing trends for the same time periods. I used MS EXCEL to determine the 21-year trends for the PSMSL global tide gauge data. I started with the most recent 21-year period, 1981 to 2001, and worked my way back in time through to the earliest period, 1855 to 1875. Oh, by the way, PSMSL stands for Permanent Service for Mean Sea Level[1], and the satellite-era data are available from the University of Colorado Sea Level Research Group[2].

"Ready for this? he asked. "Please scroll forward to the next graph, Anna."

I did. The curve showing the 21-year trends of the tide gauge-based data had large swings in those trends, over multidecadal periods. And, for one period, well before my time, the 21-year trends even exceeded the trend of the satellite-based sea level data. Dad's sea level trend graph is included below as my Exhibit 02.

Dad explained the graph, "First notice the units of the y-axis. They're the rate of change in sea level in millimeters per year, for the 21-year periods, not simply height in millimeters."

"Gotcha."

"The horizontal dashed maroon line is the 21-year linear trend of the satellite-based global sea level data, it serves as the reference, and it's at 3.2 mm/year for the period of 1993 to 2013. The blue solid curve is the 21-year trends of the long-term PSMSL tide gauge data in 1-year increments, and the trend for a given period is shown at the last year of each of those 21-year periods."

Exhibit 02

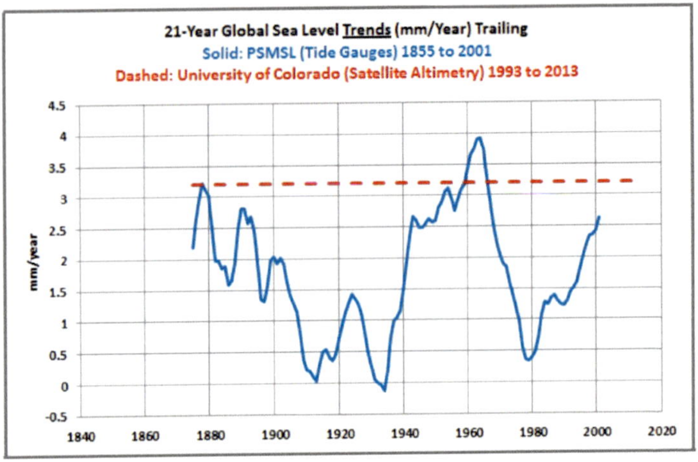

"Okay, I understand what you've done."

"As shown, the 21-year period for the tide-gauge data ending in 1878 had a trend that was comparable to the trend of the 21-year satellite-based data ending in 2013. And the 21-year rates of sea level rise ending in much of the 1960s were greater than the much-more-recent satellite-based data. Better said, the 21-year rates of sea level rise ending in much of the 1960s were greater than the supposedly CO_2-driven satellite-based data. In fact, sea levels for the 21-year period ending in 1964 rose at a noticeably higher rate than was experienced during the satellite era of 1993 to 2013."

Dad then turned his eyes to me and asked the questions I'd known he would as soon as I saw his graph. "If greenhouse gas emissions are what drive the rate of rising sea levels, how is it, Anna, that sea levels rose at a rate that was comparable to the satellite-era rate way back in the late 1800s, long before mankind's emissions of carbon dioxide are said to have been a problem?"

I remained silent.

"And how is it," he continued, "that sea levels rose faster in the mid-20th century than they have recently, and that was from the mid-1940s to the mid-1960s when anthropogenic carbon dioxide emissions and atmospheric levels were noticeably lower?"

I swallowed hard. "It appears the multidecadal rate of sea level rise can vary without the help of CO_2. And, it definitely suggests that the current rate of sea level rise is well within the realm of natural variability."

Staring at his steepled fingers, Dad said, "It could be argued then that the climate science community still *cannot* truly differentiate between human-induced and natural sea level rise." He sat looking at me silently.

Avoiding his eyes, I handed Dad's tablet computer back to him. Then I remembered the much-touted climate model hindcasts and forecasts. Brightening, I slid forward to the edge of the couch. "I think there's a hole in your argument, Dad. You forgot about the climate models! Climate models show that recent sea levels would've risen at a far slower pace had it not been for our greenhouse gas emissions, just like they show that global surface temperatures could not have risen since the mid-1970s without anthropogenic greenhouse gases."

Dad smiled "Thank you, Anna. You've now opened the door to topics I've hoped we'd get to since our discussion began. And back we go to your initial question."

"Dad, why are you a global warming denier?" I repeated.

"I deny nothing, o' daughter of mine, who learned about CO_2-driven global warming in her youth when she was gullible and naïve. I'm a heretic. In other words, my data-based, and my climate model-based, and my global politics-based opinions on this matter are different from the masses who've been brainwashed by the mainstream media. More to the point, I've studied this for decades, looking on it with an open mind as I do most things in life. My open-mindedness, as I search for answers, assures that I have no confirmation biases, no preconceptions, no prejudices. And what I've found leads me to my very firm understandings that differ from yours and every other propaganda-indoctrinated person."

My Dad was not attempting to mislead me by claiming to be open-minded. He has clearly shown me throughout my life—time after time and topic after topic—that he has no confirmation biases, no preconceptions, no prejudices. He has always explained to me how he arrived at his understandings. These are traits I've tried to follow as an adult, but with global warming and climate change I seem to have failed.

"Can we start with your data-based and climate model-based opinions?" I asked.

"Sure. You're a nerd, just like me, so we can save the politics for last. But—great big but—you need to understand the global politics behind the global warming movement if you're going to fully comprehend what's been going on with the science for the past 3 decades or more. Why don't we start with the models? Then we can discuss the data. Are you aware that surface temperature data strongly suggest there *is* a natural explanation for the surface warming the planet has experienced since the early 1980s?"

"No. And it will take a lot to convince me."

"Don't worry. You're allowed to be skeptical of me. Rest assured, though, I'll explain in fine detail and show you the data that support my understanding that the surface warming since the early 1980s happened naturally. But first, let's look at how poorly climate models simulate Earth's climate."

He rose from the couch, grabbed his pipe from the desk, and said, "If you'll give me a few moments to collect my thoughts, I'm gonna head to the garage and have a quick smoke."

"Be my guest."

CLIMATE MODELS ARE NOT SIMULATING EARTH'S CLIMATE

Dad then left me to my thoughts. From my seat before the blazing and crackling hearth, I could hear the incessant pounding of the rain against the windows as it whiplashed horizontally in sheets. This prompted my memories of an oft-repeated assertion by climate scientists before, during, and after major storms, regardless of whether they were tropical cyclones or blizzards hitting somewhere around the globe. The repeated assertion is that a warming world leads to more evaporation from soil and, more importantly, from all bodies of water, which leads to more moisture in the atmosphere and, in turn, leads to more rain or snow per storm. I could see no possible way for Dad to escape that logic.

As I ran that thought and many others around in my mind, footfalls resonated up the back uncarpeted stairwell. Dad soon joined me without fanfare, emptied the remaining ashen contents of his pipe into a covered ashtray and returned to his perch beside me.

As if he had been reading my mind, Dad asked, "What climate-related puzzle have you devised in my absence, sweetie?"

With a smile and a wink, I replied, "Your mindreading is improving, Dad."

"No mindreading involved. That was a likely question based on logical thought progressions. Well, hit me with your conundrum."

"Have you been watching the original *Star Trek* recently? You're sounding like Spock again."

"Nope."

"Okay, answer this for me. With this storm, I expect to see a climate scientist or two in the news saying that CO_2-driven global warming causes more moisture to evaporate from soils, oceans, lakes, etc., and that the added moisture in the atmosphere makes precipitation increase, making every storm worse as the planet warms. With your skepticism, do you agree with that? It's sort of a common-sense byproduct of global warming, don't you think?"

To my initial surprise, Dad agreed with the basic reasoning. "Those claims about rainfall rate do sound extremely logical, now, don't they? The same logic should apply to global precipitation, shouldn't it?"

"Sounds absolutely right," I agreed.

"I investigated the climate models used for the latest IPCC report. The outputs of those models are kept in an archive called the Coupled Model Intercomparison Project Phase 5, or CMIP5[3]. The average of the outputs of those models is available to the public online for free at a website run by the Royal Netherlands Meteorological Institute, KNMI, called the KNMI Climate Explorer[4]. That website also has lots of data from research agencies around the globe: global temperature, precipitation, cloud cover, sea ice, and lots more. And you can select the coordinates of the part of the globe you want to examine. It's a great website. Great tool. Thank you, KNMI.

"Anyway, the average of the model simulations of global precipitation shows a positive trend since the late 1970s. The trend is slightly positive—slightly, slightly positive—but it *is* positive."

Pleased, I crowed, "So the climate scientists are right!" Then I thought about the sea level data Dad had showed me, so I added, "At least about this byproduct of global warming."

"In climate models, yes. But the number-crunching computer simulations in those models are not the real world. Far from it. Very far from it. For that we need to look at data-based measurements made by rain gauges and satellites so that we include precipitation over land *and* oceans. That way we're comparing apples to apples with the models."

Quickly navigating through several pages on his tablet computer, a triumphant expression appeared on his face. I wasn't getting a good feeling about this.

"Found it," he said. "Here we have a comparison graph I prepared a couple of years ago." He handed me the tablet computer and continued, "The dashed red curve and trend line represent the average of the climate model simulations of global precipitation."

I've included Dad's model-data comparison graph of global precipitation below as my Exhibit 03.

Exhibit 03

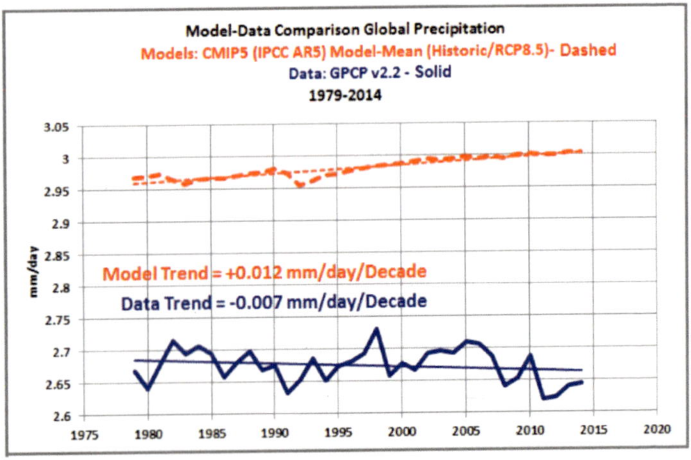

Source of Data and Climate Model Outputs = KNMI Climate Explorer

Dad continued, "As noted a few moments ago, the dozens of models included in the average are from the CMIP5 archive, which was used by the IPCC (Intergovernmental Panel on Climate Change) for their most recent assessment report. I used the average of the model outputs because it best represents the consensus—the groupthink—of the climate modeling-groups around the globe."

"And the disappointing data in the solid blue curve, where's that from?" I asked then noted, "They showed a slightly negative trend, contradicting the climate models."

Dad stated, "The dataset is made up of satellite-based and rain gauge-based data. It's from NOAA (the National Oceanic and Atmospheric Administration). It's their Global Precipitation Climatology Project data version 2.2, also known as GPCP v2.2 from NOAA[5]. It's also available on the KNMI Climate Explorer. The global precipitation data show a slight decrease, roughly about -0.9% based on the linear trend over those very recent 35 years, but the models show we should have seen a very slight *increase* in global precipitation over the same period, roughly +1.4% based on the linear trend."

He studied my face before he continued. "The biggest disparity, however, and I mean big, is in the amount of precipitation shown in the models. The average of the models is showing more than 10% too much precipitation. That might not sound like a lot, but think about this."

"What's that?"

"Well, according to the modeled worst-case

scenario, global precipitation is only supposed to increase about 7% by the year 2100."

Then the realization hit me. "But the difference between the modeled and observed global precipitation in recent decades is 10%, and that's greater than the expected 7% worst-case increase by the year 2100."

"Bingo."

Then I realized the significance of the models showing an increase in precipitation over the past few decades while the data showed a decrease. My heart sank. "So in their sermons, when climate scientists serve as experts for the mainstream media, they might not mention it, but they might be referring to the *virtual* worlds of their computer simulations and not the real world." I inadvertently sighed with disgust as I handed the tablet computer back to my father.

"Indeed. And because the data contradict their sermons, it's likely they're referring to the models. But based on that graph, we have to consider a much more important implication of that big difference between the observed data and the models. All of the excessive precipitation indicates there may be way too much water vapor in the number-crunched make-believe atmospheres in the models. Water vapor is a greenhouse gas, the most abundant of all the greenhouse gases. Logic then dictates that the higher the water vapor, the greater the rate of warming in the models. That is, if their water vapor guess is too high, so too is their guess about global warming."

"In simpler terms then, Dad, it sounds like you're

saying the climate scientists are sneakily forcing speedier and greater global warming in the models by using that excess water vapor."

"That's exactly what it looks like. Seems to me, the models didn't generate warming fast enough to satisfy the politicians. So the funding-hungry toadies, known to the rest of the world as climate scientists, added more water vapor to the atmospheric mix to force the politically expedient global warming results."

"Well at least the models do a good job of simulating global temperatures."

"Do they, Anna? Are you basing that on your own research, or are you parroting some nonsense from some scientist or the mainstream media?"

I flinched with his use of the word *parroting*, because it injured my pride. But that's actually what I was doing, parroting what I'd read online. So I pushed it aside and said softly, "I suspect you're going to show me I'm wrong, Dad."

"Not wrong. Just misinformed. Sorry if I upset you."

"No reason to be sorry, Dad. I *was* parroting what I'd heard on the news or elsewhere online."

I was becoming increasingly uncomfortable with the thoughts that there was a lot of misinformation out there from climate scientists, even more uncomfortable with the thoughts that gullible and naïve people were then, in turn, being misled by mainstream media, led down the garden path of a fairytale world.

"Well, Anna, are you ready for a quick look at how poorly climate models simulate global surface temperatures?" Dad asked as he rummaged through files on his computer.

"Ready when you are."

"While I'm searching, I had a thought a couple of years ago about the IPCC reports, well at least the summaries for policymakers. The bottom lines don't really change at all from report to report. Better said, the bottom lines haven't changed from the first IPCC report back in 1991 through to the most recent 5th Assessment Report in 2014. That is, they all say the same things: As a result of anthropogenic greenhouse gas emissions, primarily CO_2, global surfaces will warm, sea levels will rise, and droughts and floods will get worse. But they can't tell us where and during what decades the floods and droughts will occur so that people can be prepared. On the more-technical side, from the first assessment report back in 1991 to the most recent one, the climate scientists have not narrowed down Earth's climate sensitivity to a doubling of atmospheric CO_2 levels. They're as clueless now as they were 26 years ago."

"Sounds right, Dad, now that you've said it. With the tens of billions of dollars spent annually on climate science, there hasn't been a lot of, if any, worthwhile new information from that group in decades, only more and more alarmism."

"And with that as my lead-in," Dad said, but let it linger for a moment before he stated, "Of course you understand that our planet Earth was poorly named. About seventy percent of its surface is covered by Ocean."

"Yes, I believe it was Arthur C Clarke who wrote

'How inappropriate to call this planet Earth when clearly it is Ocean'."

"You have a great memory, Anna. So one would think climate scientists might have looked to the surface temperature data for the oceans when checking whether their models were simulating surface temperatures properly. Obviously, they never bothered."

Dad handed me his computer yet again. The graph it displayed is shown in my Exhibit 04, furnished below.

Exhibit 04

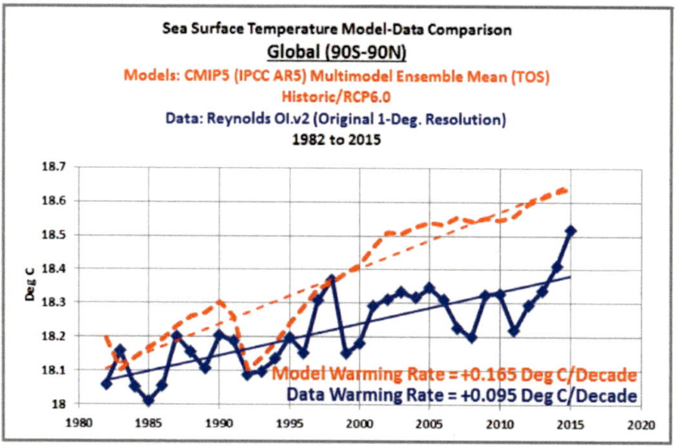

Source of Data and Climate Model Outputs = KNMI Climate Explorer

"As you can see, Anna, the groupthink of the modeling groups show that from 1982 to 2015 the surface of the global oceans from pole to pole should have warmed at a rate of 0.165-deg C/decade, but the satellite-enhanced data show a warming rate of less than 0.1-deg C/decade."

"And the surface temperatures in the models are too high," I said, before I noted, "The data end in 2015. Why do they begin in 1982?"

"Don't worry, Anna. I haven't cherry picked the start year to exaggerate how poorly the models perform. That dataset is *satellite-enhanced*, and it's from NOAA. It's their Reynolds Optimum Interpolated Sea Surface Temperature, version 2, a.k.a. Reynolds OI.v2[6]. That dataset starts in November 1981, so the first full year is 1982. I created that graph a couple of years ago when I was studying climate model outputs. I wasn't using the info for any project. It was only for my personal info. If you want to update that graph to include more recent data, be my guest. The results won't be significantly different. The trend of the data may have increased slightly, about 0.01 deg C per decade, if that much.

"But I do have to warn you," he continued. "Note the latitudes. They're 90 South to 90 North. That is, I've included the polar oceans with their *sea ice* in that graph. The data supplier and the modelers may account for sea ice differently, so you'd be better off excluding the polar oceans."

"Gotcha."

"If you would, please scroll ahead one time for the same model-data comparison, but excluding the polar oceans and their sea ice."

I performed as asked, and the graph that appeared is included below as my Exhibit 05. I observed, "Oh, I see, you've limited the latitudes to sixty degrees North and South, thereby excluding the Arctic and Southern Oceans. The difference in the warming rates between the models and the data is slightly greater, Dad, when you take away the polar oceans."

Exhibit 05

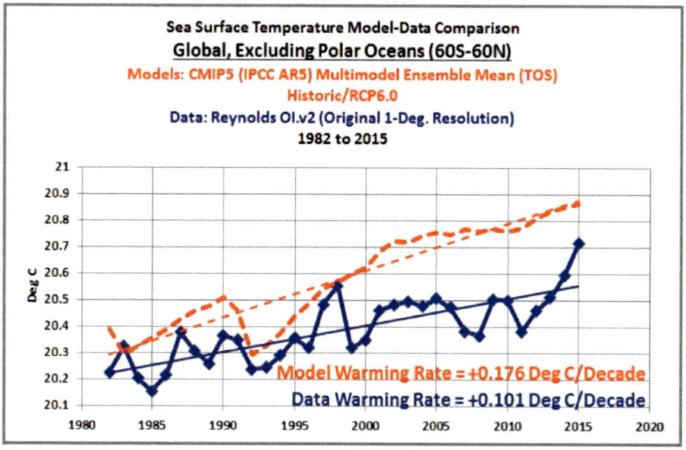

Source of Data and Climate Model Outputs = KNMI Climate Explorer

"Agreed, honey, but the warming rates of the models and data have not changed significantly. The models are showing too much warming by roughly 70 to 75%, either way. Imagine if the models used by investment firms performed that poorly."

I chuckled. "I'd fire them in an instant and find some other firm to help me invest my money."

"So would I, so would I. Sadly, we don't have another climate science community to go to. We're stuck with those pitiful excuses for scientists who've been forever tainted by politicians and their agendas." Dad, with an uncharacteristic fire in his eyes, growled.

"Nice growl."

His mood lightened. "Thanks."

"And again, the climate modelers' guesses at sea surface temperatures are too high," I observed. "Based on the linear trend lines, the models start off too warm by a little less than 0.1-deg C and end too warm by about 0.3-deg C."

Then the disparity between modeled and observed precipitation rates (my Exhibit 1) came to mind. "Something seems very skewed in the models, Dad. Your earlier model-data comparison of global precipitation showed the models were producing precipitation that was 10% too great over a similar time period. You then remarked how the consensus of the modeling groups showed global precipitation only increasing 7% by the year 2100. And that brings me to a question."

"Fire at will."

"According to the climate models—you know, the ones used by the IPCC for bad guesses at future global warming—how much more are Earth's ocean surfaces supposed to warm by the end of this century?"

"Based on the worst-case scenario, the climate models used by the IPCC for their most recent report indicate global sea surfaces, excluding the polar oceans, will warm roughly 3-deg C by the year 2100."

"That's basically what I was thinking when I said something seems very skewed," I said. "If a 3-deg C rise in global sea surface temperatures will increase global precipitation by about 7%, then the models are really screwed up if they show global precipitation rates now that are 10% higher than the observations with modeled sea surface temperatures that are roughly only 0.2-deg C higher than the observations. That's way too big a difference between modeled and observed global precipitation for the tiny difference between modelled and observed global sea surface temperatures."

"*Brava*, Anna. When we look at climate model outputs and compare them to the real world, the models, as you say, seem very skewed—very skewed indeed or very *screwed up* as you also said—and we don't have to look at them very closely to see how skewed—screwed up—they truly are."

Dad stared at the cheerfully roaring fire for a moment, then added, "Always keep in mind, for the past few decades, climate models really only exist to support political agendas and to help global warming profiteers like Al Gore and that electric car guy. Climate models have not existed to gain any worthwhile insights into our warming world."

"How evil!"

"It truly is."

"Hey, Dad, you were talking earlier about excluding the polar oceans from the global sea surface temperature comparisons due to possible differences in how the data providers and modelers account for sea ice. I was wondering then if you'd also run model-data comparisons for sea ice."

"Wonder no longer. May I have my computer back?" he asked. I complied, and he began scrolling through a series of graphs with blue and dashed red curves like the two model-data comparisons we've already discussed for global sea surface temperatures. He explained all of those additional graphs, "When I prepared the two *global* sea surface temperature graphs we just examined, I investigated how poorly the models performed for each ocean basin—the Atlantic, Pacific, Indian, Arctic and Southern Oceans—and at different latitudes and hemispheres. I'm running through all of those graphs now, and, as you can see"—I scooted over to get a better view—"they continue to show models that perform horrendously when compared to real world data. But that would be information overload right now. It may take me a...oh, here we are, a model-data comparison of northern hemisphere sea ice area."

"Thanks, Dad. I really want to see it. I always wondered about a news report I saw maybe ten years ago. I was in high school, I think. In it a couple of climate scientists were talking about how the decline in Arctic sea ice was worse than projected by the latest and greatest climate models. I believe the catch phrase was 'worse than we thought'."

"I remember that crap," said Dad, a twinkle in his eye.

"It struck me in a way then that the scientists may not have been expecting." I chuckled. "And it was, 'worse than we thought'? That doesn't speak very highly of their climate models if their predictions are off so soon."

"Right you are...again, sweetie."

Dad was silent while he surfed the web, but then he suddenly said, "Change of topics. Consider this study from a few years ago that confirmed what many people have always thought: far and away more human deaths are caused by cold than heat. The paper Mortality risk attributable to high and low ambient temperature: a multicountry observational study[7], by Gasparrini, et al., was published in a medical journal, *The Lancet*, not a science journal, in July, 2015. There was an article in *USA Today* about it. Ah, there it is. Study: Cold kills 20 times more people than heat[8]. Would you like to read it, Anna?"

"I remember that story. I wondered then why global warming was supposed to be bad, if cold kills more people than heat. Back then I figured it meant that precipitation and sea level were the key problems with global warming. But, you've already straightened me out on those two aspects."

"Excellent. Since we've gone off on tangents, did you know the UN ran an online survey to determine what people around the globe thought were the biggest problems facing their world today?"

"Nope."

"Can you guess where climate change ranked with all of the global respondents?"

"It's not first, is it?"

"No, climate change ranked last."

"We wouldn't know that by how the mainstream media treats climate change."

The survey results Dad referred to can be found at UN's My World 2015 data[9].

Dad then returned to the topic at hand. "Why don't we take a look at a model-data comparison graph of Northern Hemisphere sea ice area?" He handed me the tablet computer with the graph displayed. I've included it below as my Exhibit 06.

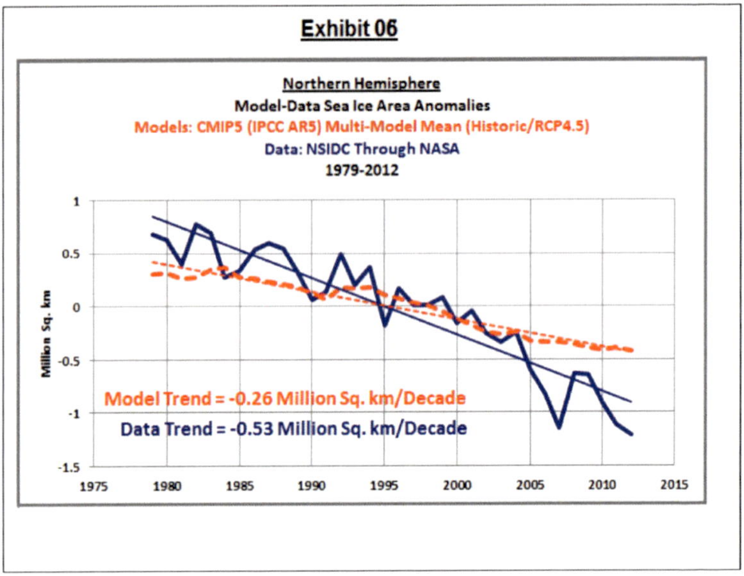

"I didn't know it was that bad," I said. "The Northern Hemisphere sea ice in the models declined at *half* the observed rate. That's a horrible performance by the models."

"Scroll ahead, and you'll find an even bigger problem with the modeled sea ice in the Southern Hemisphere."

So I did. And there it was plain as day. The consensus of the climate modeling groups guessed that the sea ice area in the Southern Hemisphere would *decrease* from 1979 to 2012 if CO_2 was the driver that caused changes in sea ice area. But the satellite observations-based data show an *increase* in area. The graph prepared by Dad for this metric is presented below as my Exhibit 07.

Exhibit 07

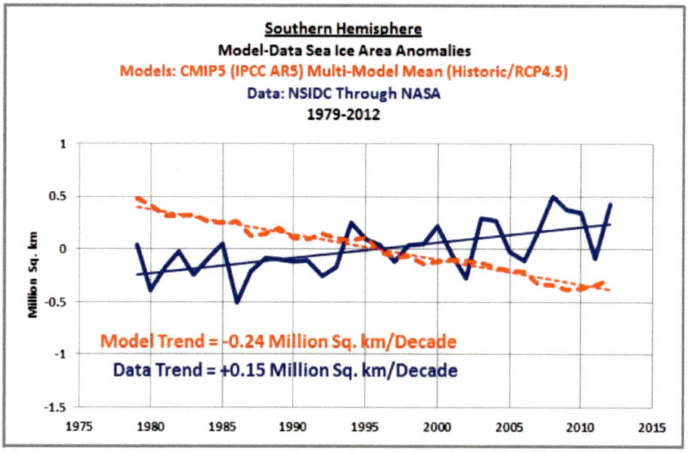

"That is incredibly poor performance on the part of the models," I said.

"Indeed. One might suspect the model programmers, a.k.a. climate scientists, had *not* even bothered to examine the data to see whether the sea ice in the Southern Hemisphere should be increasing or decreasing in surface area. Regardless, they should be embarrassed with that horrendous performance.

"*And* yet they have the gall to continue to hold out their hands begging for more taxpayer dollars for ever more time on mainframe computers so they can run their wretched number-crunching software that in no way represents Earth's climate."

Suddenly, I became aware that I was hearing something that struck me as odd. After listening for a moment, I realized what it was. "The rain stopped," I announced happily.

"And the sun is beginning to break through the clouds," Dad replied.

"You know what's sad, Dad? Whenever it rains, whenever it's warmer or colder than normal, whenever it's cloudy or sunny, whenever it snows, whenever there's a flood or a drought somewhere, or a blizzard or hurricane, I'm so brainwashed that all I think about is that humans are responsible for whatever whenever. When the brainwashing began, I wondered *if* we were responsible. No longer, though. With the constant browbeating, until this visit with you right now, I was convinced we *are* responsible. No questions asked."

My Dad reached over and took my hand in his. He seemed to know I wanted to vent more, and he waited for me to continue.

I did. "But the propaganda had a horrible effect on me, Dad. I don't enjoy weather anymore. I'd always loved weather, with all of its varieties, its strengths, its violence, its tranquility. Those are awe-inspiring splendors of Mother Nature, her gifts to us. And I no longer enjoy them. And now, after what we just talked about, that makes me incredibly sad, Dad. Propaganda has stolen those daily joys from me."

"Agreed, honey. Weather in all its varieties is awe-inspiring. Oh yeah, and brainwashed is definitely the right word. You've been brainwashed by politicians, by their toadies the climate scientists, by mainstream media, by enviro-profiteers, and by all that gloom and doom garbage that's online."

He stood. "Let's take a break for a couple of minutes, let you quiet down. I need bring in some more firewood."

"Sounds great. I'll make us some more tea?"

Afterwards, now settled back on the couch and our teas refreshed, Dad said, "Let's return to modeled global temperatures, shall we?"

"Please."

"First, here's a graph I prepared not long ago, during the build-up to the wealth-distributing Paris Agreement. They claimed the goal of that treaty is to limit global warming to 2-deg C above preindustrial temperatures. I wanted to explore when the climate models projected we would reach that point with the IPCC's worst-case scenario. And I included data to see where we stood."

News headlines and some illustrations flashed through my mind. "I seem to recall there were plenty of graphs at that time, showing we were about half-way there."

"Once again, your memories are correct." Dad handed me his tablet computer again. "Note how soon the models say we may reach that 2-deg C limit."

The model-data graph Dad presented is included below as my Exhibit 08.

"Five years?" I chuckled. "Then again, at the other end of the spectrum, we reach that threshold about 40 years from now."

"The rectum of the spectrum."

Exhibit 08

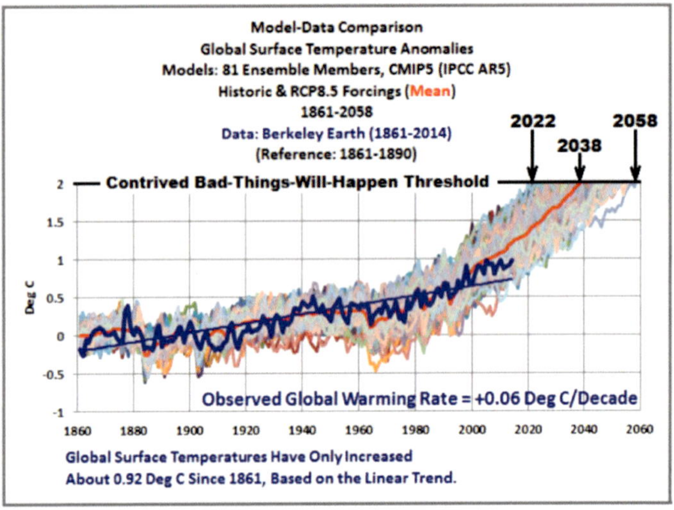

I laughed harder with his juvenile humor then said, "Let's see, the model mean is the red curve, so the consensus shows us reaching that 2-deg limit in 2038, about two decades from now."

"That is, of course, if you find the crystal ball-like climate models credible sources for prognostications. And, based on the linear trend of the data, it took more than 150 years for global surface temperatures to rise less than 1-deg C, yet the consensus of the modeling groups are expecting global surface temperatures to skyrocket the other degree in only two decades."

"Ah, we're forgetting something, Dad. There was a secondary goal of the Paris Agreement." I pulled out my smartphone and searched. "Here we are '...pursuing efforts to limit the temperature increase to 1.5-deg C above pre-industrial levels…'". I started to laugh again, looking at the graph.

Dad joined me with a belly laugh then said, "Now, you're getting the hang of this. The models are so bad they're laughable. According to some of the models, we should've already passed that 1.5-deg C limit, passed it a couple of years ago."

I could only smirk, so Dad filled in the dead air, explaining, "The models in that graph are once again made up of those included in the CMIP5 archive, which was used by the IPCC for their most-recent report. And the future forcings are for the worst-case scenario. The global surface temperature anomaly data is from the Berkeley Earth[10] group, not one of the government agencies like NOAA or the UK Met Office."

"Dad, all surface temperature datasets are very similar. And there's a fundamental reason for that. They're all based on the same source data. The only differences between them are caused by the adjustments each supplier makes to the raw data."

"Agreed. Nice summary. Back to this graph: The Berkeley Earth dataset is spatially complete, so we are comparing apples to apples with the models. The anomalies for this graph are referenced to the first 30 years of the models, 1861 to 1890. That's as close as we can come to pre-industrial times, given the 1861 start year of some models."

"Anomalies." I admitted, "When I first heard a climate scientist on the news use the phrase *global temperature anomalies*, it made me think of the words unusual and strange. In other words, that something about the temperature was unusual or strange."

"But it sounds like you have a different understanding now."

"Yes I do. Temperature anomalies are simply an easier way to say the deviations or departures from the average temperatures during a selected time period, usually a 30-year period."

"Right you are. And the person presenting the data can select the 30-year period for the anomalies, whatever 30-year period they want. My objective with this graph was to show the change in temperature from pre-industrial times, or as close as I could get to preindustrial times. So I used the first 30 years of the climate model average."

"Anomalies can be very confusing to newcomers...and it was when I was new to this."

"Okay, do you have any other thoughts about this model-data comparison?"

"Overlooking the obvious shortcomings, the long-term model hindcasts look pretty good, to my eye," I noted.

"Hindcasts, where the model programmers tune their software to mimic the historical data."

"With all the models presented as a thick cluster, they seem to perform reasonably well," I said. "The group of model outputs contains the annual and long-term variations the data, and the model mean doesn't stray very far from the data until recently."

"Climate scientists love to present their models as a multi-model array, many times as a big-wide grey cloud surrounding the data. They do that for precisely the reasons you've given. That big wide strip includes wiggles from 81 individual computer model runs so that it encompasses the short- and long-term wiggles of the data.

"But take another look," he continued. "Look closely at the data and *model mean*, not the big cloud, from the early 1910s to the early 1940s. The data show a rate of global warming that's much greater, much-much greater, than that shown by the model mean. In other words, the consensus of the modeling groups indicates that global temperatures *should have* warmed at a much lower rate. That is, if the global surface temperatures were dictated by CO_2 and other forcings that drive the models, surface temperatures should not have risen as fast as they had from the 1910s to the 1940s. Said again, with all of their Ouija-board conjectures and speculations about what causes global warming, the modelers can't explain why global surfaces warmed as fast as they did from the 1910s to the 1940s."

"But the warming rates of the model mean and the data align much better from the mid-1970s to 2000. They actually look great during that period, and that's where my eyes are drawn, to where the models perform well," I said, still wondering if the climate models might have some meaningful predictive skill.

Dad sighed. "The models are *tuned* to perform well in that time period. Point of fact, the IPCC used the term 'tuning knobs' in their second assessment report, referring to the software inputs that drive the number-crunching models."

"But the 1970s to 2000 is the period the scientists point to when they're showing the public their models have matched the past. And of course the public then assumes the models are valuable tools for predicting the future."

"If they tune the models to perform well during that period, they can't use that period to validate the models."

"But they do, Dad."

"I understand, and I think you know the point I'm trying to make."

"Let's see if I do. Because they tune the models to perform well during the latter part of the 20th Century, they're only showing how well they can tune the models, not how well the models actually perform at simulating past and future global temperatures."

"Precisely. But that's what they do with their show and tell, Anna, all the while constantly misinforming the public. Now let's take another look at climate model performance. But this time I'll show it in a way you'll never see presented by climate scientists." He paused. "Because it highlights the abysmal failings of climate models at simulating the longer-term changes in global surface temperatures."

Dad then asked, "If you would, please scroll through a few more graphs. You're looking for a graph titled Model-Data Comparison using 30-year, 360-month, trends, or something to that effect. Let me rephrase that; the graph is displaying 30-year *trends* in modeled and observed global surface temperatures, not simply global surface temperature anomalies. And the *trailing* in the title block means the trends are shown at the last month of each 30-year period."

"This one?" I asked.

"Yes, ma'am."

It was the next graph in that file. I've included it below as my Exhibit 09.

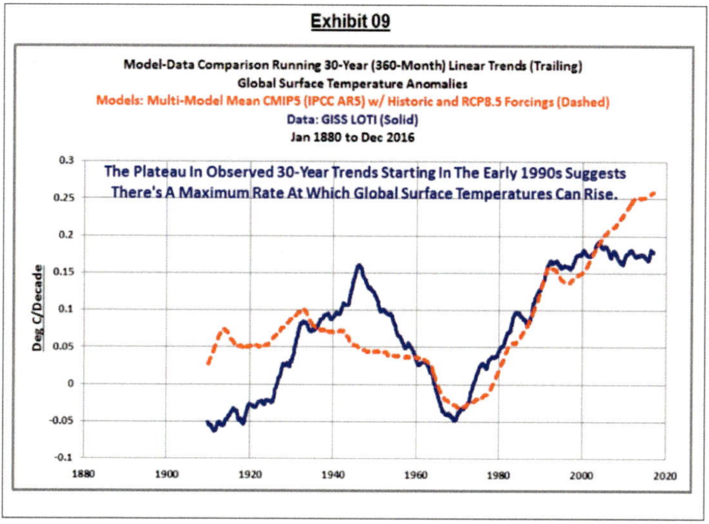

I observed, "Your statement that the models are tuned to the latter part of the instrument temperature record looks to be confirmed by this model-data presentation."

"I'll confirm later that the models are tuned to that period, using the Second Assessment Report from the IPCC."

"But looking at this graph, the models perform horribly prior to the 30-year period ending in the 1960s. And the divergence is striking between the models and data after what looks to be near 2005. Oh, and for the observations-based global temperature anomaly data in this graph, you've used the Land-Ocean Temperature Index (LOTI) data from the NASA Goddard Institute of Space Studies."

"Yup on all counts. Did you see my reminder note that reads 'The plateau in the 30-year trends starting in the early 1990s suggests there's a maximum rate at which global surface temperatures can rise'?"

"Couldn't miss it," I said. "And it's plain as day that the recent plateau in the global warming rate isn't too much higher than the earlier peak rate ending about 1945. That peak ending in the mid-1940s is the real standout when compared to the models. It shows that surface temperatures rose at a 30-year rate that was more than three times faster than hindcast by the models."

"Correct. If the variations in Earth's surface temperatures were dictated only by the numerical representations of the anthropogenic and natural forcings that are used to drive the models, then the surface temperatures should only have risen at a rate of 0.05 deg C/decade for the 30-year period ending in the mid-1940s."

"But they warmed at a rate that was more than three times faster," I said. "That suggests *at least* two-thirds of the more-recent warming spell since the mid-1970s could be natural as well. And it also strongly suggests, in turn, that climate model sensitivity to a doubling of atmospheric CO_2 is way too high, three times too high."

"Anna, that assumes the models are simulating Earth's climate. But, as we've already clearly seen with the earlier model-data comparisons—the sea ice graphs, *and* the sea surface temperature graphs, *and* the global precipitation graph—climate models are *not* simulating Earth's climate. Climate models, as they existed in the past and as they exist today, have no scientific value, no scientific value whatsoever. None. Nada. Zip. Bupkis. Because they're not simulating Earth's climate, I can think of no possible reason why we should give a damn about how climate responds to CO_2 on some fantasy number-crunched virtual planet that bears no relationship to Earth."

I let my father's words replay a few times in my mind.

After a few moments of silence, he continued, "Now look at how poorly the models simulate the global surface temperature trends toward the early portion of the graph. For the 30-year periods ending in the 1910s and early 1920s, the data showed *global cooling* at low rates was occurring, but the models show that *global warming* should have been occurring at low rates. The models couldn't simulate global cooling when it occurred early on. And, as a result, they couldn't simulate the warming as global surface temperatures rebounded."

The poor performance of the models before the 1960s forced me to accept some things I never would've suspected when I left my apartment that morning. "They tune their models to mimic the warming in the latter part of the 20th Century, but by doing so the modelers and climate science community in general can't explain the early warming. So the additional early warming has to have occurred naturally. But the early warming rate is similar to the rate of the latter—supposedly CO_2-

caused—warming. In other words, that graph clearly shows that climate scientists still cannot differentiate between human-induced and natural global warming."

"That's blatantly obvious in that graph."

"And...if we were to take into account that the models aren't simulating Earth's climate, *and* take into account the uncertainties of the data, then we really have to say the recent warming is well within the realm of natural variability."

"Bingo, my brilliant daughter."

"Now I have a much better understanding of your skepticism, Dad."

"Thank you."

"You've been perspicuous, as always, with your explanations, Dad."

"I try to be as clear as I can."

"Darn, you even know what *perspicuous* means."

"What's that, your word of the day?"

I nodded and said, "Was a few days ago."

"Good word. I ran into perspicuity a couple of years ago."

We were drifting away from the topic of those deplorable climate models, and I was starting to get hungry, so I checked my watch. We had been happily chatting away for hours.

Dad must also have had hunger pangs. "We can put this aside if you like, and continue it after dinner."

"That sounds great. Thank you for the education today, Dad. I never would have suspected climate models performed so bad at simulating Earth's climate. The all-trusting public has been led to believe climate science is *settled*—what a joke—all the while, the number-crunching climate models are simulating the virtual climate on some fantasy science fictional planet."

"One of those fantasy *Star Trek* planets."

I needed to free myself of my frustrations with climate science so I added to Dad's *Star Trek* reference. "And it's a Klingon planet, with the climate scientists clinging on to their worthless models."

"As we've seen, Anna, the current generation of climate models performs horrendously at simulating Earth's past and recent climate, on a global basis. Believe it or don't, they're even worse when trying to simulate climate on a regional basis."

"And regionally is where we need realistic forecasts."

"Ah, forecasts, the magic word. The climate science community doesn't make forecasts. They simulate future climates on fantasy planets based on projections of future greenhouse gas emissions. So they use the term *climate projection* instead of forecast."

"And it's not a semantics thing, Dad. I watched a climate scientist correct a newscaster once on TV, and the scientist was adamant about their not providing forecasts."

My father said, "And to think, the number 5 in CMIP5 designates these are the fifth generation of climate models to be archived for research needs. You might think it's tough to imagine how horribly those models performed back in 1995 when the politicians first claimed the scientific community could now see mankind's influence on climate."

"How's about we look at that after dinner?" I asked.

"Agreed. Maybe tomorrow, though, since tonight after dinner we want to look at how the instrument temperature record indicates the warming since the early 1980s occurred naturally. And we should look at the pathetic global politics before we look at the early models because they go hand in hand.

"But let me restate something you said a few moments ago, Anna. In reality, if climate scientists were honest with themselves, they would *still* be saying they could *not* differentiate between natural and human-induced global warming."

"I agree, Dad. To sum up what I've learned today about those crappy climate models used for the reports by the Intergovernmental Panel on Climate Change, they do not properly simulate past or current global surface temperatures, or global precipitation, or hemispheric sea ice area. And you've shown me, based solely on data, that there is nothing unusual about the recent rate of sea level rise. It was surpassed in the middle of the 20th Century when greenhouse gases were at much lower levels and emission rates were lower."

Dad and I did break for a few hours, me with a newfound skepticism of the climate model-based pseudoscience that supports the hypothesis of human-induced global warming.

SATELLITE-ERA SEA SURFACE TEMPERATURE RECORDS STRONGLY SUGGEST THE SURFACES OF THE GLOBAL OCEANS WARMED NATURALLY

SINCE THE EARLY 1980S

That night, after the dinner dishes were stowed in the dishwasher, and after we had consumed our bowls of ice cream, we were seated again at the dining room table, having more tea. Dad explained his statement that data indicate that the global warming since the early 1980s has occurred naturally. It was a very detailed explanation, but once again his explanation was clear and he supported it with data.

He began, "Have you ever heard of El Niño and La Niña events, Anna?"

"Most news stories about El Niños call them unusual warming events in the eastern tropical Pacific Ocean, off the west coast of South America. They're the cause of the huge upward spikes we see in the global surface temperature graphs."

"They are much more than just warming events, and, further, regardless of what the numbskull science reporters say, there's nothing unusual about them." Dad shook his head disgustedly. "Magnificent would be a better word. Here are the facts. El Niño events occur every two to seven years. El Niños are the most-amazing, *and* the most powerful, weather events ever devised by Mother Nature. How powerful? El Niños are often kick-started by series of tropical storms in the western tropical Pacific.

"How's about we start with an overview about how they cause long-term global warming and do it naturally. Ready?"

"Ever since I got home, Dad."

"Okay, hold on for a few minutes while I get the globe from your bedroom." Dad returned with a globe I've had sitting on a bookshelf since I was in grade school. He gave it an enthusiastic spin and sat beside me, before placing it on the table between us. He stopped the globe's spinning and turned it so the Pacific Ocean faced us. From my viewpoint, the Pacific Ocean, with very little land, was all I could see. It was a great reminder of just how massive the Pacific Ocean was.

As if reading my mind, Dad said, "As you can see, Anna, the tropical Pacific Ocean stretches almost halfway around the globe."

"From Indonesia in the west to South America in the east," I confirmed, tracing my index finger along the equator. "The Pacific Ocean is humungous!"

"Indeed. Now, the surface waters in the eastern tropical Pacific are normally cooler than the waters in the western portion. And there's a simple explanation. The eastern boundary currents along the coasts of North and South America return cool waters from the extratropics to the tropics." He traced his index finger southward along the west coasts of North and Central Americas then traced it northward along the west coast of South America. "During normal conditions, the trade winds push those cool waters from the eastern tropical Pacific all the way halfway around the globe across the tropical Pacific, under the warm tropical sun, pushing it ever westward until all of that sun-warmed water runs into Indonesia and Australia." While he said the last sentence, he pretended to push water from east to west all the way across the tropical Pacific, from the west coast of South America at Ecuador, to Indonesia and Australia.

He then traced his finger along the equator in the eastern tropical Pacific, as he said, "In the eastern equatorial Pacific, east of the dateline, the trade winds also cause cool water from deep below the surface to be drawn *to* the surface in a process called upwelling. That cool water is also warmed by the sun as it travels to the west, pushed by the trade winds, almost halfway around the globe in the tropical Pacific."

"Okay, got that."

"The western boundary currents carry those sunlight-warmed waters toward the poles where those poleward-traveling waters release the sunlight-created heat to the atmosphere." As he spoke, he continued his demonstration, finger-tracing poleward in the South Pacific, east of Australia, then switching hemispheres and tracing the North Pacific east of Asia.

"Got that too, Dad. You're describing ocean circulation."

"Bingo. However, at the same time, a lot of the sunlight-warmed water accumulates in the western tropical Pacific in what's called the Pacific Warm Pool." With his finger, my father roughly circled an area east of Indonesia and north of Australia. "So, imagine a pool of water about the size of the United States, one thousand feet deep."

"That's a huge chunk of warm water."

"Yup. Everything about the Pacific Ocean is huge.

"The massive currents that carry that sunlight-warmed water from east to west in the tropical Pacific are called the North and South Equatorial Currents."

"Makes sense."

"Along the equator, there's also a much smaller current that travels in the opposite direction, from west to east, called the Equatorial Countercurrent. And below it there's another west-to-east current that runs below the surface called the Cromwell Current, also known as the Pacific Equatorial Undercurrent. Those eastward-moving surface and subsurface currents are normally carrying *much* less water than the westward-moving North and South Equatorial Currents."

"I've got you so far, Dad. And you've said sunlight-warmed a bunch of times."

"Just setting the stage and reinforcing that fact."

His use of the word *normal* in his explanations prompted me to interject, "But something happens that causes changes in those normal conditions."

"Bingo!" He smiled broadly. "A westerly wind burst, sometimes caused by a tropical storm—or two of them straddling the equator—upsets the balance and sends a huge pulse of warm water from west to east along the equator. Keep in mind that it's normally warmer in the western tropical Pacific than it is in the eastern portion. That pulse of warm water moves along the surface, but most of it is carried below the surface along the Cromwell Current. If there are enough of those westerly wind bursts in the western tropical Pacific one year, and if enough warm water travels east at and below the surface to raise the normally cooler *surface* temperatures in the central and eastern equatorial Pacific by more than 0.5 deg C, then El Niño *conditions* are said to be taking place."

"That's pretty cool, Dad."

"Nope, pretty warm, Dear."

I groaned, with a smile.

"And if the surface temperatures *stay* elevated in the central and eastern equatorial Pacific for about half a year, then a full-fledged *official* El Niño event is said to have taken place."

"Gotcha. El Niño *conditions* means the sea surface temperature *anomalies* in the eastern equatorial Pacific have risen above a 0.5 deg C threshold, and if they stay elevated for about half a year, then an *official* El Niño event is taking place."

"And El Niño events typically peak during the boreal winter. The westerly wind bursts start much earlier in the year. It takes a couple of months for the warm water from the West Pacific Warm Pool to travel eastward toward South America. Now keep in mind that it takes multiple westerly wind bursts to send enough warm water eastward to cause an El Niño. So the birth of an El Niño by Mother Nature takes about a half a year, sometimes longer, sometimes shorter. Oh yeah, El Niños typically start in one year and end in the next. Thus you'll see them expressed, for example, as the 1997/98 El Niño."

"It must be fascinating to watch El Niños form."

"NOAA and Australia's Bureau of Meteorology (BOM) monitor lots of atmospheric and oceanic variables in the tropical and equatorial Pacific, so you can actually watch the slow birth of an El Niño online as it occurs over many months.

"Now, let me clarify something. In an El Niño, a *monstrous* volume of warm water traveled from the West Pacific Warm Pool to the eastern side of the tropical Pacific, where it's normally cooler. No new warm water was created. It was just relocated *from* the Western Pacific Warm Pool, where it's normally warmer, *to* the eastern equatorial Pacific, where it's normally cooler. *And* the warm water that had been below the surface, traveling from west to east along the subsurface Cromwell Current, it gets upwelled to the surface in the east."

"Okay, I understand."

"And during *very* strong El Niño events, so much warm water is carried east that most of the warm water, above and below the surface, is now in the eastern tropical Pacific instead of where it normally is in the western tropical Pacific. That is, the surface and subsurface conditions—normally warmer water in the west and cooler in the east—have traded places."

"That's a lot more detailed and helpful than the simple explanation we hear on the news. Then, because the news media call it an unusual *warming* event, people like me who don't understand the process get mixed up and think the El Niño is caused by global warming."

"When just the opposite is the case. An El Niño can cause a *naturally* occurring, sunlight-fueled, long-term global warming."

"So *that* is the natural causation you were talking about," I said nodding enthusiastically, "El Niño!"

"Right you are. But before I get into that in more detail, I need to clarify something. Because of the process called upwelling, during the El Niño, there is more warm water than normal spread across the entire tropical Pacific, sometimes as far east as the coast of South America. With all of that warmer-than-normal water spread across the eastern tropical Pacific, much more evaporation is taking place there. So the tropical Pacific is releasing monumental amounts of sunlight-created heat to the atmosphere." My father paused, gave me an attention-grabbing look, and said, "But, not all of the warm water that has traveled east is cooled all the way back to normal by evaporation. So, and this is very important, the renewed trade winds push all of that leftover warm all the way back across the tropical Pacific, being warmed a second time under the tropical sun—let me repeat that, being warmed a *second* time under the tropical sun—before being sent poleward by the western boundary currents."

"In a hot water heating system," I said, thinking of my current and Dad's former work, "that would be like recirculating warm water a second time through the boiler, heating it more, before pumping it out to the heating coils and baseboard radiation."

"Yes!" Dad chuckled. "That's precisely what I thought when I first realized this."

"So how did you identify this? Did you run into a scientific paper about this?"

"Nope. The climate science community looks at an El Niño as noise on the global surface temperature record, not as a natural cause of global warming. That is, the so-called scientists completely discount El Niños, when in reality El Niños are naturally occurring, sunlight-fueled sources of global warming. *My* realization came after *I* divided sea surface temperature data into three logical subsets:

1. the North Atlantic Ocean because it has a well-known mode of long-term natural variability called the Atlantic Multidecadal Oscillation;
2. the entire Eastern Pacific Ocean because it showed no surface warming for more than 30 years; and
3. that left the rest of the global oceans—the South Atlantic, Indian and West Pacific Oceans—and they clearly show that they warmed in steps."

"They warmed in steps?"

"Yup, you can't miss them. And once I understood how El Niño and La Niña events really worked, the steps made a lot of sense, *a lot* of sense. Excuse me while I go grab my computer."

"I'll clear the table while you're gone."

"That's okay, sweetie. Rest, relax," he said, jumping up. He gathered the dishes, and hustled out to the kitchen to deposit them with a clatter in the sink before dashing upstairs to his study.

Alone for a few minutes with my thoughts about naturally occurring global warming, it was only a few moments before a thought struck me. I voiced it, calling out to my Dad, "The climate scientists have been claiming for more than two decades that global warming has been caused by mankind's emissions of CO_2. If it turns out to have happened naturally, they are going to look to the world like blithering, incompetent fools...and what you called them, funding-hungry toadies to politicians."

Dad called back from upstairs, "And, of course, the first scientists to openly discuss this would be labeled deniers, and their funding would be cut by the politicians."

Dad's footfalls in the hallway above me then on the stairs heralded his return to the dining room. After resuming his seat beside me, he handed me the tablet computer and said, "Just in case you doubted my veracity about the lack of surface warming on the eastern Pacific Ocean, the entire eastern Pacific Ocean, take a look."

Exhibit 10, below, is the graph my father furnished me. It includes data for the sea surface temperatures of the East Pacific Ocean, from November 1981, the start of that satellite-enhanced sea surface temperature dataset, to December 2013. The trend line was calculated and created by MS EXCEL.

"I didn't doubt you, Dad. But Wow! This certainly confirms what you told me. A warming rate of +0.01 deg C/decade. That's basically no warming for more than 30 years."

"Guess how much the climate models used by the IPCC for their predictions of gloom and doom say the surface of the East Pacific should've warmed during that time."

Exhibit 10

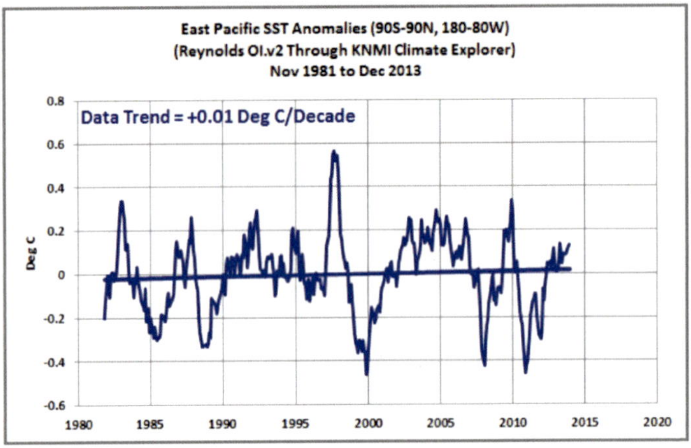

Source of Data = KNMI Climate Explorer

"No idea," I replied. He paused for effect, so I broke the silence, "Well, how much?"

"The best guess of the programmers of those useless number-crunching climate models is, if human CO2 emissions controlled surface temperatures, over those 3-plus decades, the sea surfaces of the East Pacific Ocean, from pole to pole, and from the dateline to the west coasts of the Americas, should have warmed about 0.45 deg C. But the data indicate the sea surface temperatures did *not* do that. The sea surface temperature *data*—the satellite-enhanced observations—say that it only warmed...wait for it...0.03 deg C in 30 years."

"In other words," I said, "CO_2–driven climate science says the surface temperatures of the eastern half of the entire Pacific Ocean should have risen almost 0.5 deg C from 1982 to 2013, but data show there was basically no warming there. None."

"Yup."

"Let's see if I've got this. The big upward spikes show the impacts of El Niños on the surface temps of the East Pacific and the downward spikes show the impacts of La Niñas."

"Yes, however, a La Niña is not as simple as the typical weather reporter makes it out to be. La Niña events are not the opposite of El Niño events. A La Niña is the normal state of tropical Pacific but on steroids.

"When the trade winds resume their normal direction after an El Niño, they blow all of the leftover warm water from the El Niño back westward under the tropical sun, so there is a greater-than-normal temperature difference between the Eastern and Western Tropical Pacific. The greater temperature difference causes the trade winds to blow harder, and that causes more upwelling of cool waters from below the surface of the equatorial East Pacific. We can see that reflected on the graph as the downward spikes, the stalactite-looking ones.

"If you do any online research about this, you'll often see discussions of three states of the tropical Pacific: Normal, El Niño, and La Niña. Because an El Niño is the truly *anomalous* state of the tropical Pacific, I, like many El Niño researchers, believe there are only two states: El Niño and other, where other is comprised of Normal and hyper-normal, a.k.a. La Niña."

I turned the globe until the Eastern Pacific Ocean was front and center to us. "The East Pacific is one giant chunk of water. And there had been no warming there for more than 30 years."

"That chunk covers about 30% of the surface of the global oceans," my father confirmed.

I added, "And because the oceans cover 70% of the entire globe, that means the East Pacific covers a little more than 20% of the surface of the entire globe, while land only covers 30%. And according to the linear trend of the data, the East Pacific showed no surface warming for more than 30 years."

"Another great summary, sweetheart. And now for the evidence that strong El Niño events were responsible for the surface warming of more than 50% of the global ocean surfaces." Dad reached over and scrolled ahead to the next graph in the series.

I've included that graph as my Exhibit 11, below.

He said, "Here we have an annotated graph of the sea surface temperature anomalies of the South Atlantic, Indian and West Pacific Oceans from late in 1981 to December 2013."

"No way," I cried. "The warming there really did happen in steps. You're right. You can't miss them."

"That's what I saw, and guess what. I could find no studies by the so-called climate scientists that addressed that curiosity. No studies whatsoever. No mention in the IPCC's fifth assessment report, either. If *I've* plotted it and seen it, then *they* have to know it exists."

Exhibit 11

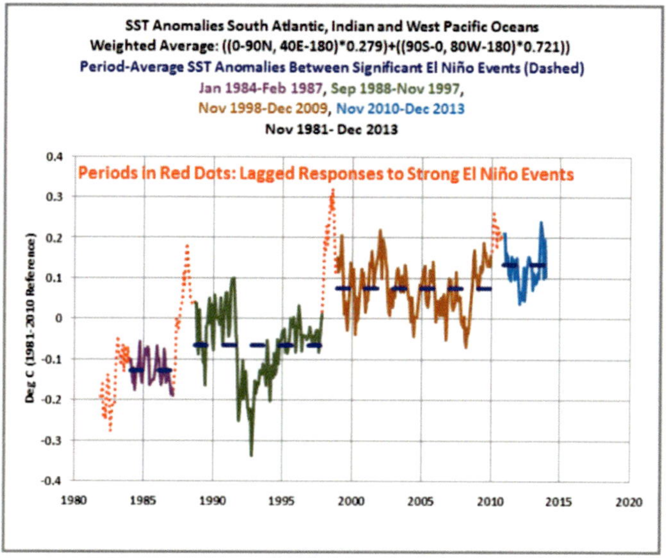

Source of Data = KNMI Climate Explorer

Dad began pointing out things on the graph. "First, even though I've broken the data into shorter time periods of different colors, the wiggly, step-like data run continuously from November 1981 to December 2013. And that was the full extent of that satellite-enhanced sea surface temperature dataset back then when I created that graph. The dotted red portions cover the time periods when the strong El Niño events of 1981/82 and 1986/87 and 1997/98 and 2010/11 impacted this part of the global oceans. To properly determine the timing of those dotted red months, I used NOAA's Oceanic Nino Index[11], which lists the months of *official* El Niño events. Then I shifted those periods by six months because there is a time lag from when the El Niños were occurring in the East Pacific to when these parts of the global oceans responded to them. The horizontal dashed lines in blue are the average sea surface temperatures for the periods between the El Niño events."

"Okay, I've got you. Now let me guess: that big dip and rebound in the data in the early 1990s, that's the effect of the volcanic eruption of Mount Pinatubo…the result of all of those sun-blocking volcanic aerosols being ejected high into the atmosphere."

"Right you are, Anna."

"Something's missing from that graph," I said.

Dad motioned with his hand toward the screen. "Be my guest." He smiled.

"Where are the stalactite-looking dips and rebounds on the graph that represent the impacts of the *La Niña* events?"

"They're simply not there. That is, the data are silently shouting, *La Niña events are not the opposite of El Niño events*, no matter how many times climate scientists and the mainstream media tell us they are!"

# # #

I was going into overload with so much El Niño and La Niña info. In the months that passed since we had this discussion, Dad has answered my emailed questions and clarified the amazing properties, functions and processes that take place in the tropical Pacific, and in the atmosphere above it, during El Niño and La Niña events. In the tropical Pacific, the ocean and atmosphere above it are coupled, and their interactions are fascinating. The strength and direction of the trade winds impact the sea surface temperatures, and the sea surface temperatures impact the strength and direction of the trade winds. Thus, they are said to be coupled.

He also explained how La Niñas serve to replenish the ocean heat released during an El Niño. And that fact makes the utmost sense. How this works is, during a La Niña, the sea surface temperatures of the tropical Pacific are lower than normal. Because of the lower surface temperatures, there is less evaporation than normal from the surface of the tropical Pacific. Less evaporation causes less cloud cover above it. With fewer clouds, more sunlight can enter into and warm the waters of the tropical Pacific to depth, thus replenishing the heat released during the prior El Niño. Depending on the weather conditions in the tropical Pacific at that time, the La Niña can replenish part of the heat lost during the prior El Niño or all of it. In some cases, a La Niña can create more heat than was lost in the prior El Niño, and that creates more ocean heat for the next El Niño.

The ocean heat content data of the tropical Pacific, and a scientific paper, confirmed his understandings that the warm water for the 1997/98 El Niño was created in months, not decades, during the rather weak La Niña of 1995/96. Keep in mind that the 1997/98 El Niño permanently raised the surface temperatures of South Atlantic, Indian and West

Pacific Oceans by approximately 0.15 deg C. See my Exhibit 11 above.

# # #

"You said something earlier that has me a little confused, Dad, that an El Niño does not *create* any of the warmer ocean water, but, rather, that an El Nino just relocates the warm water and that the relocated warm water raises global surface temperatures."

"Okay, first off, the La Niña, through a reduction in cloud cover, allows more sunlight than normal to reach into the tropical Pacific, so the La Niña creates the warm water, not an El Nino.

"Second thing: The global surface temperature datasets are made up of two types of data. For the surface *air* temperature on *land*, the data come from weather stations that are sampling air temperatures roughly six feet above the surface. For the oceans, they use the surface temperatures of ocean waters, a metric called sea surface temperature, not the temperatures of the air near the ocean surface, a metric called Marine Air Temperature. So, because the oceans cover more than twice as much of the Earth's surface than land, global surface temperature data are made up mostly of sea surface temperature data.

"Now, do you recall our discussion of the Pacific Warm Pool in the western tropical Pacific?"

I nodded and then said, "The Pacific Warm Pool is located east of Indonesia and north of Australia. It's about the size of the United States. And the sunlight-created warm water accumulates there to depths of about 1000 feet. It accumulates there because the trade winds have pushed the waters from east to west across the entire tropical Pacific, or about halfway around the world, under the tropical sun, until it reaches Indonesia, and those land masses restrict its further transport west."

"Very good, Anna. So before the El Niño, there's a huge volume of warm water pooled up there in the western tropical Pacific, with most of it below the surface. But because it's *below* the surface, it's not included in the *surface* temperature record. Let me repeat that, because it's important. Before an El Niño, there is a huge volume of water below the surface of the West Pacific Warm Pool that will be used by the El Niño, but, because the warm water is below the surface, it's not included in the surface temperature record. Then, during an El Niño, that subsurface warm water is relocated to the east and upwelled to the surface in an area where its normally cooler, so now there is more warm water included in the surface temperature record."

"Oh, I get it, Dad." I stood and paced before the fireplace. "Before the El Niño, most of the warm water in the West Pacific Warm Pool was below the surface of the western tropical Pacific and excluded from the surface temperature records. And then, during the El Niño, that warm water is carried east and spread across the surface in an area where it's normally cooler. Let me rephrase that to clarify it. All of that warm subsurface water that used to be excluded from the surface temperature data before the El Nino is now included in it during and after the El Niño. That's like an instant global *surface* warming during an El Niño, a surface warming that's created by sunlight-warmed water. Then, after the El Niño, the warm surface water that's left over from the El Niño is pushed back to the west by the trade winds, all the way across the tropical Pacific, almost halfway around the world, and it's being heated by the tropical sun *a second time*."

"I believe you've got it now, Anna." Dad smiled. "And that warm water that's been heated twice by the tropical sun is carried around the globe by ocean currents, warming the surfaces of Western Pacific, the Indian Ocean and the South Atlantic in turn."

"And it creates those upwards steps in the surface temperatures of the South Atlantic, Indian and West Pacific Oceans that's shown in your graph, Dad." (See my Exhibit 11 above.)

He nodded and then said, "One last note about El Niño and La Niña events acting together as a process. Together they act as a chaotic, naturally occurring, sunlight-fueled, recharge-discharge oscillator...with El Niños acting as the discharge phase and La Niñas acting as the recharge phase. But La Niñas also help to redistribute the leftover warm waters from the El Niños...water that has been heated a second time after the El Niño under the tropical sun."

"That's definitely a better description than you hear from the TV weatherperson. Let me see if I've got it."

"Give it a shot."

I did, pausing occasionally. "As a combined process, El Niños and La Niñas behave as a chaotic...naturally occurring...sunlight-fueled...global warming-producing...recharge-discharge oscillator...with El Niños acting as the discharge phase and La Niñas acting as the recharge phase. But La Niñas also help to redistribute the leftover warm waters from the El Niños...water that has been heated a second time...in the tropical Pacific...after the El Niño...under the tropical sun."

"Brava, Anna." My father applauded for a few seconds, then scrolled through a few more graphs on his computer, talking while he searched. "So, we've got the East Pacific, which hadn't shown any warming for more than 30 years. And we've got the big subset made up of the South Atlantic, Indian and West Pacific Oceans, which, the data show, clearly warms naturally in steps in response to strong El Niño events, but it has only warmed—naturally, I might add—at a rate of slightly less than 0.1 Deg C/decade. And, finally, we have the North Atlantic, with its own mode of natural variability called the Atlantic Multidecadal Oscillation, many times referred to as the AMO. And guess what. The North Atlantic has been in a naturally occurring warming phase for more than 30 years, where it's surface is warming faster than the rest of the oceans, at a rate of a little more than 0.2 Deg C/decade."

Dad then handed me the tablet computer. The sea surface temperature anomalies for those three regions—that makeup all of the global oceans—were presented in a graph, which I've included below as my Exhibit 12.

"So, dear daughter of mine, how well do you think the climate models used by the IPCC simulated the trends in sea surface temperature for those three regions? And, of course, keep in mind that those three regions make up the global oceans."

Exhibit 12

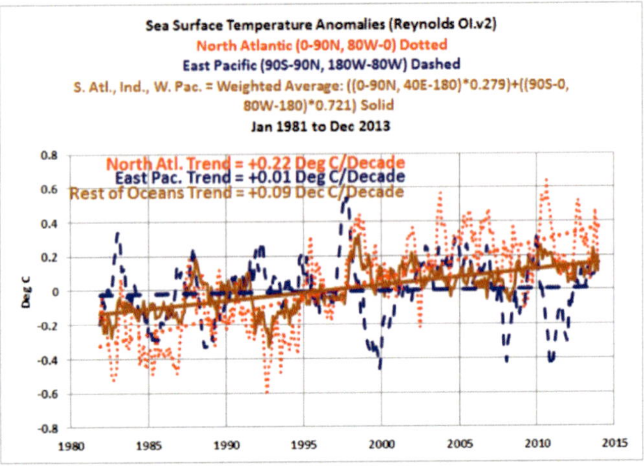

Source of Data = KNMI Climate Explorer

I had to chuckle. "I suspect you're gonna show me the models performed like poo."

"And right you are. Poo they be. The models do a reasonably good job simulating the short-term warming rate of the North Atlantic. But, they performed abysmally in the East Pacific, saying there should have been lots of warming where no surface warming existed for more than 30 years. And for the biggest subset, the South Atlantic, Indian and West Pacific Oceans, the models show the surfaces there should be warming almost 80% faster than they have. Keep in mind the models are telling us how much the ocean surfaces should have warmed if, big if, they were warmed by the manmade greenhouse gases that drive the models. But we've already discussed how the surfaces there are warmed in responses to naturally occurring and sunlight-fueled El Niño events."

"You've mentioned something a couple of times, Dad. What's the Atlantic Multidecadal Oscillation?"

"It actually takes place in the North Atlantic. It's a mode of natural variability—caused by naturally occurring ocean processes—that allows the surfaces of the North Atlantic to enhance or suppress global surface warming over periods of three to eight decades."

"Interesting. There are naturally occurring ocean processes taking place in the North Atlantic that can enhance or suppress global warming, surface warming that is."

"Now, when I say enhance global surface warming, I mean those processes can cause global warming to accelerate for periods of three to eight decades."

"Then suppress means that those processes can cause global warming to decelerate. Let me guess. Since the 1970s, the North Atlantic has been in the phase where it enhances global warming."

"Right you are, Anna. And climate scientists blame that temporary acceleration in global warming on our increased emissions of greenhouse gases, primarily CO2. And to reiterate a point, the sea surface temperature data show the warming occurred naturally, where it occurred."

"Thanks."

"There was a paper about a decade ago which

said the supposedly correct way to present the Atlantic Multidecadal Oscillation with data was to subtract the global sea surface temperature anomalies, long-term, from the sea surface temperature anomalies of the North Atlantic. If memory serves, the paper was by Trenberth and Shea." He started to surf online. "Let me check... Here it is, Trenberth and Shea (2006) Atlantic hurricanes and natural variability in 2005[12]. In it, they proposed to subtract global sea surface temperature anomalies excluding the polar oceans, 60 south to 60 north, from the sea surface temperature anomalies from the corresponding part of the North Atlantic, using 0 to 60 north and 80 west to 0, which excludes the appropriate portion of the Arctic Ocean. So for the next graph, I prepared the data-derived Atlantic Multidecadal Oscillation, using a long-term sea surface temperature dataset from the UK Met Office called HADISST[13]. And I compared it to the modelers' guesses of it, using the same coordinates from the models stored in the CMIP5 archive."

"Hold on, Dad. Did you just say those scientists said that the Atlantic Multidecadal Oscillation was better determined by subtracting global sea surface temperature data from North Atlantic sea surface temperature data, using a long-term sea surface temperature dataset?"

"Yup. I think I know where you're going, but please go ahead."

"The global data includes the North Atlantic so they are not getting a true representation of the additional variability of the surface temperatures of the North Atlantic. They're reducing the variations by subtracting North Atlantic data from itself. Granted, the North Atlantic portion of the global data is reduced in impact because of the data from the rest of the oceans, but it's still part of the global data." I shook my head slowly in a frustrated disbelief. "If those climate scientists were truly interested in determining the additional variability of the North Atlantic versus the rest of the global oceans, then they'd take the North Atlantic sea surface temperature data and subtract from it the sea surface temperature data from the rest of the world, not all of the global oceans."

"You are right, once again, Anna. I suspect that the proper method you described didn't give them the results they wanted, so they took a shortcut and found what they wanted."

I was becoming disgusted. "Only in climate science."

"Back to the model-data comparison of the Atlantic Multidecadal Oscillation, using the method prescribed by Trenberth and Shea for both the models and the data." He handed me his computer and continued, "And as you can see, the consensus of the models do not show anywhere near the same variations, in time or in magnitude. Also note that the data and model outputs have been smoothed with centered 121-month running-average filters, which is commonly used in the presentation of the Atlantic Multidecadal Oscillation."

Dad's model-data comparison graph for the Atlantic Multidecadal Oscillation is my Exhibit 14 below.

"Wow, not even close," I said, shaking my head in disbelief.

Exhibit 14

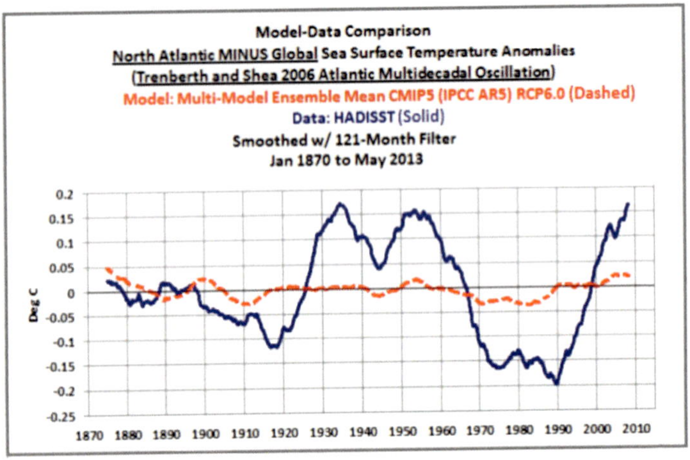

Source of Data and Model Outputs = KNMI Climate Explorer

"Another thing about the multi-model mean," Dad said. "It averages out all of the potential yearly, decadal, and multidecadal variations from the individual model runs, which are chaotic and out of sync with one another. That, of course, assumes the individual models do, in fact, produce decadal and multidecadal variations. So, because all the model outputs are averaged to get the model mean, they eliminate the noise from the individual models, and that leaves us with something special: the forced component of the models. That is, by averaging the outputs of all of the models, we're eliminating all of the junk noise from the models and leaving ourselves with the best guess by all of the modeling groups for how the Atlantic Multidecadal Oscillation should vary if those variations were only caused by the numerical representations of the manmade greenhouse gases, etc. that are used to drive the models. In other words, *if*, big if, the additional multidecadal variations in the sea surface temperatures of the North Atlantic were dictated by manmade greenhouse gases and the other forcings used to drive the models, then those variations in sea surface temperatures would resemble the red dashed curve. And as you can plainly see, it's clear there are no similarities between the actual Atlantic Multidecadal Oscillation and the modelers' groupthink about it."

I concluded something at that moment. "The results from the computer programmers are so poor, that one cannot conclude much of anything from climate models...other than that the programmers' assumptions about CO_2 are wrong. To put it in simpler terms: the modelers' guesses are not based on reality. Thus, climate model outputs are, at best, *useless*, but, at worst, they're *highly misleading* for those who do not understand them, like the general public."

"With that graph, it's blatantly obvious that there are naturally occurring processes taking place in the North Atlantic that enhance and suppress global warming that are not accounted for when the funding leeches, who call themselves scientists, make their unfounded, nonsensical proclamations about Earth's sensitivity to a doubling of carbon dioxide levels."

"It's scary when you think about it, Dad. Climate models have been elevated to religious icon status by climate scientists, by politicians, and by enviro-profiteers—who make money off solar, wind, electric cars, speaking engagements, etc.—and in turn by the average true believer."

"And, it seems that climate scientists can no longer grasp that their models are not the real world."

"Beyond that, there's no way the average person can understand that explanations from models are fictional at best, misleading and misinforming at worst. And to make it even worse, if someone who actually studied the models was to tell the average person that the models have been programmed to provide results that are intended to scare the public into behaving as the politicians want them to behave, then that person who studied the models and found their flaws is considered an aluminum-foil-hat-wearing conspiracy theorist."

"Sounds like you've been talking to your Uncle Marty."

"Want me to make you an aluminum foil hat?" I asked with a broad smile.

Dad returned my smile. "How about an aluminum tri-corner hat like Captain Jack Sparrow's?"

I laughed aloud. "Really?"

He chuckled. "No thanks. But *you'll* be getting one from Uncle Marty for your birthday, if he finds out about your new-found skepticism."

"No way I'm gonna talk to *him* about global warming ever again. Not ever again."

My father shifted our conversation back to climate models, but on the future forecast side, not the hindcasts of the past and present as we had been discussing. "Back to the model-based prognostications: the worst-case scenario used by the modelers as inputs for future greenhouse gas emissions, etc., is something called RCP8.5."

"I've seen that in lots of articles. Isn't that the one that's often described as the *business-as-usual scenario*?"

"That's exactly what I wanted to discuss. It's not business as usual. It's totally unrealistic."

"Why don't I find that surprising anymore?"

"RCP8.5 assumes no technological progress in our sources of energy as we march through the 21st Century. In fact, it assumes that coal once again becomes the primary source of energy. How ridiculous! Coal as the primary source of energy in the year 2100; imagine that!"

"So you're telling me that, not only are the modelers not simulating Earth's climate, they're using totally unrealistic inputs of future sources of energy, inflating future greenhouse gas emissions. In other words, the modelers are just wasting time on mainframe computers, consuming the same electricity as small towns with each of those mainframe computers, for totally unrealistic future scenarios taking place on fantasy planets."

"That brings us to one of the problems with policy-driven science, Anna. The politicians have to scare the public in order to gain public support for the high costs of the political agendas, and the scientists are more than willing to genuflect to their deep-pocket masters, like the good puppets they are."

"With that, I'll be imagining Pinocchio's nose growing anytime I hear a climate scientist yacking about human-induced global warming."

Dad smiled and suppressed something he wanted to add but was having second thoughts about it.

So I returned to the topic at hand: "Suppose the global warming scare evaporated, Dad. Climate scientists would be out of a job. Considering the trash they're presenting as science, that would be a pretty good idea."

"And instead of paying them unemployment benefits, we could put them to work building dykes and levees around low-lying cities for the rising sea levels that are gonna occur regardless of our greenhouse gas emissions."

"That would be a better use of the supposed scientists' time. Instead of their playing on computers and flying to exotic locations for climate conferences, let them do some manual labor for once."

We both paused for a few moments.

I was burning out from the hours of eye-opening conversations, so I tried to bring them to an end, at least for the night. "I think I've had enough of global warming and climate change for today. And since we've been talking about science fiction for so long, have you seen Season 2 of *Stranger Things*?"

"No, honey, but I did watch Season 1 a couple of weeks ago to refresh myself. And I'm up for Season 2, if you are."

"Let's do it."

"First, how about a summary about the sea surface temperatures of those three regions?"

A little disappointed we weren't moving immediately to the family room to watch *Stranger Things*, I replied, "The surface of the Eastern Pacific Ocean hadn't warmed in more than 30 years, based on satellite-enhanced sea surface temperature data. The data for the large subset, which covers more than half of the surface of the global oceans—made up of the South Atlantic, Indian, and West Pacific Oceans—clearly shows the surfaces there warmed naturally in response to strong El Niño events. Then there's the North Atlantic, which has a mode of natural variability called the Atlantic Multidecadal Oscillation, and the naturally occurring AMO caused the surfaces of the North Atlantic to warm faster than the rest of the global oceans, which also warmed naturally, where they warmed."

"So, basically, since 1982, the surfaces of the global oceans warmed naturally."

"Did we somehow exclude the polar oceans, Dad?"

"Nope, the data for the three subsets we discussed included their portions of the Arctic and Southern Oceans. Check the coordinates in the title blocks of those graphs, Anna."

"What about land surfaces, the other 30% of the surface of the globe?"

"Land surfaces warm primarily in response to the warming ocean surfaces. There was a study about a decade ago that made that revelation." He searched the web, muttering "Compo, I think," then quickly exclaimed, "Eureka! It *was* Compo and Sardeshmukh (2009) Oceanic influences on recent continental warming[14]. The abstract reads, and I quote":

> *Evidence is presented that the recent worldwide land warming has occurred largely in response to a worldwide warming of the oceans rather than as a direct response to increasing greenhouse gases (GHGs) over land. Atmospheric model simulations of the last half-century with prescribed observed ocean temperature changes, but without prescribed GHG changes, account for most of the land warming. The oceanic influence has occurred through hydrodynamic-radiative teleconnections, primarily by moistening and warming the air over land and increasing the downward longwave radiation at the surface. The oceans may themselves have warmed from a combination of natural and anthropogenic influences.*

He sighed before continuing. "The problem is, they used climate models to determine that land surfaces warm in response to warming ocean surfaces. Granted they were special models, where sea surfaces temperatures are used as inputs. But they were still climate models."

I added, "But the sea surface temperature data show that the surfaces of the oceans warmed naturally."

"Bingo. Let's go watch *Stranger Things*."

# # #

As we watched the first two episodes of season 2 of *Stranger Things*, I pondered our belief-changing exchange on the topics of global warming and climate change, during which he altered my beliefs, not vice versa. I asked myself why I hadn't fared better in the debate, especially since I've been exposed to global warming teachings—better said, propaganda—throughout my teens and twenties. The answer was, Dad had data and climate model outputs to support his understandings. Where I, for my beliefs, only had nonsensical alarmist hoopla furnished almost daily by mainstream media, by teachers in high school, by professors on college, and, more than a decade ago, by the alarmist trash found online, and by Al Gore's misleading *An Inconvenient Truth*.

Those thoughts led to memories of a news story that Dad had shown me years ago about a British judge who had ruled there were nine errors in *An Inconvenient Truth*[15], and, if teachers wished to play the movie for their classes, they first had to expose those errors to their students, or at least that's my recollection from back then.

While he watched *Stranger Things*, Dad tinkered away on his computer, updating a couple of graphs. He presented them the next day after he returned from showing that duplex in town.

# # #

Late Saturday morning, I greeted Dad at the kitchen door. “Yay, you’re home!”

He wrapped me in his arms and smiled. “Yay, so are you!”

“Any luck with the duplex?”

“They’ve made an offer the owner will definitely accept.”

“Tea to celebrate, oh teetotaler?”

He glanced at the kitchen table and my mug there. “What are you having?”

“Hot chocolate.”

“One of those, please. Let me go change into some sweats, and I’ll meet you in the man cave.”

And so we returned to our places on the couch.

“I called your Mom last night before I went to bed. Told her we were having a great time.”

“Me too. Uncle Marty’s got pneumonia, or so he claims, so she’s over with him instead of being at Gram and Gramp’s. Did you hear what he’s blaming the pneumonia on?”

“I can only imagine,” Dad replied dryly, slowly shaking his head with rolled eyes and smirk.

“If you imagined he was blaming global warming for that chilly spell last week, then you were right.”

“He blames everything on global warming. Even the two coyote pups playing in our backyard this summer were brought here by global warming.”

“I’ve still got that video on my phone that Mom sent. They were beautiful. Even now, I can’t get the smile off my face.”

Dad placed what was left of his mug of hot chocolate on the coffee table and retrieved his computer from his desk. He returned to our discussions from last night, "I updated a few graphs last night while we were watching Netflix."

"I noticed. Whatcha got?"

He handed me his tablet computer. "Here's an update of the sea surface temperature anomaly graph for that large part of the global oceans that warms in steps in response to El Niño events. There was another El Niño in 2014/15/16, so there was another upward step. Not very big like the one in response to the 1997/98 super El Niño, but it's there. Keep in mind that this graph represents more than 50% of the surface of the global oceans."

The updated graph of the sea surface temperature anomalies for the South Atlantic, Indian and West Pacific Oceans, through September 2017, is my Exhibit 15 below.

"And there's evidence right there in the data that the surfaces of the South Atlantic, Indian and West Pacific Oceans warmed naturally in response to strong El Niños," I added, beating him to the punch.

"Very good. Remember the model-data comparison graph for the Atlantic Multidecadal Oscillation?"

"Yup, for that you subtracted the global sea surface temperature anomalies from the North Atlantic sea surface temperature anomalies, doing that with both a long-term sea surface temperature dataset and the model outputs."

Exhibit 15

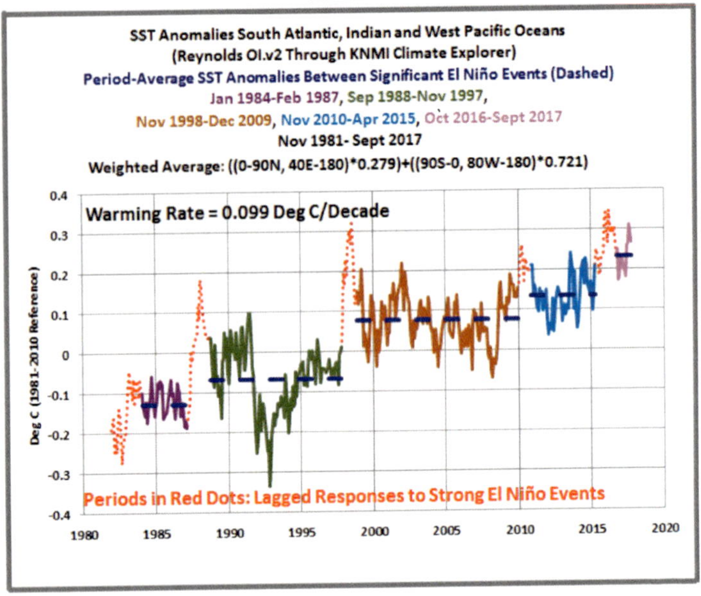

Source of Data = KNMI Climate Explorer

"And the results?" Dad looked at me, waiting.

"What is this, a quiz?" I said with a broad smile. "There is no way I would ever forget that graph. The data showed that the surfaces of the North Atlantic can warm or cool over multidecadal periods at rates that are much faster than the global oceans. In other words, the North Atlantic can suppress global warming or enhance the natural warming taking place in response to El Niño events. And because the climate models do not show those variations in time or amplitude, we have to conclude the additional variations in the sea surface temperatures of the North Atlantic have occurred naturally."

[I of course was thinking of my Exhibit 14 above.]

"Brava!" my Dad exclaimed in a way that warmed my heart, as only my Dad can.

"Scroll back a couple of graphs, please."

So I did. And the graph he prepared is shown below as my Exhibit 16.

Exhibit 16

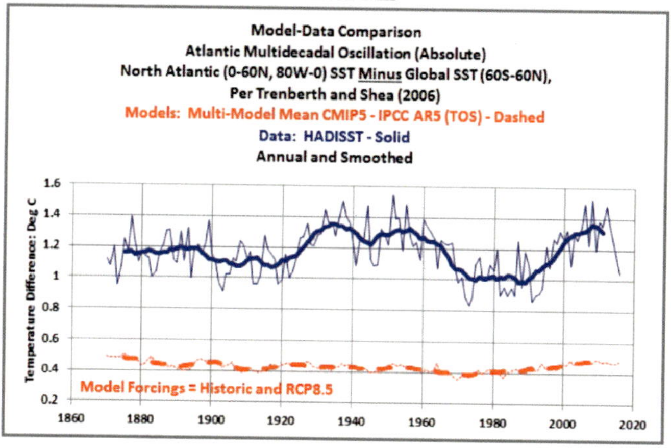

Source of Data and Model Outputs = KNMI Climate Explorer

I noted, "This one looks different than the last one."

"You're right, sweetie. This time I used annual sea surface temperatures in absolute form. I wanted to see if the modelers could simulate the observed temperature difference between the surfaces of the North Atlantic and those of the global oceans. Of course they couldn't. They're worthless."

"I can see that," I agreed, amazed not that the climate "scientists" were wrong, but *how very wrong* they were.

"The model outputs and data are in annual form and smoothed with 11-year running-average filters, centered. Oh yeah, keep in mind, to agree with the paper Trenberth and Shea (2006), I've excluded both polar oceans."

"Got that. Says so in the title block. Let's see. What haven't we re-examined?" I asked, in an effort to prompt him.

"The East Pacific. Okay, scroll ahead a couple of times. It changed a bit, but you have a general idea what it looks like."

The updated graph of the satellite-era sea surface temperature anomalies for the East Pacific Ocean, from pole to pole, is shown below in my Exhibit 17.

"Changed a bit? Holy smokes! From 1981 through 2013, the East Pacific Ocean had showed no warming, but now it shows warming at a rate of 0.07 deg C/decade. What happened? Was it a monster El Nino?"

"There was an El Niño that lasted from November 2014 to May 2016, but it didn't cause that all by itself."

"We know that manmade greenhouse gases could not have bypassed the East Pacific for more than 30 years and then all of a sudden decided to cause some warming there. So what happened?"

Exhibit 17

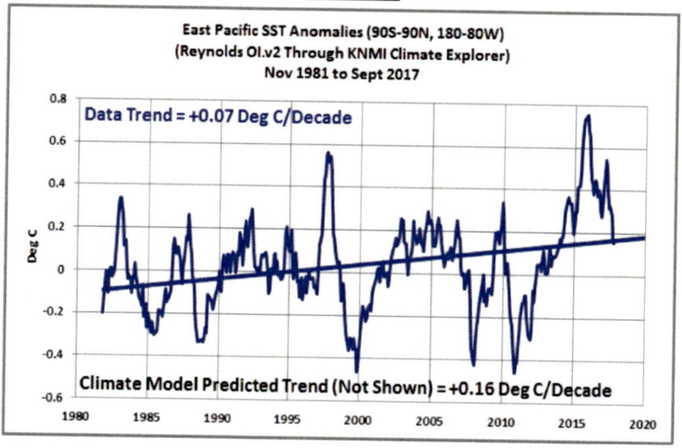

Source of Data and Model Outputs = KNMI Climate Explorer

Keep in mind the surface of the East Pacific showed little to no warming from 1982 through 2013. See Exhibit 10. Regardless, alarmists now claim the new warming trend is caused by anthropogenic greenhouse gases, when they still cannot explain the lack of surface warming that existed there for more than 30 years.

"That's a great overview, sweetie. Great overview. I'm gonna paraphrase you. For more than 30 years, the surface of the East Pacific showed no warming despite increases in atmospheric CO2 levels, totally contradicting the climate models that are based on the hypothesis of human-induced global warming. But then, when Mother Nature creates a couple of major upticks on top of one another, the alarmists claim that it's just what the climate models predicted."

"Sad."

"Okay, Anna, do you remember the alarmistproclaimed *permanent drought* in Southern California that wasn't so permanent and only lasted for a couple three years?"

"Yup, and after it ended, they were complaining about rainfall and flooding in California. Are you saying the California drought and that jump in the temperatures of the East Pacific are related?"

"Yes ma'am. They were both caused by a ridge of high pressure over the eastern extratropical North Pacific. You know, a ridge of high pressure like the TV meteorologists talk about. Well, this one lasted an extra-long time, and, as a result of that resilient ridge of high pressure, a very large pool of warm water formed there. It was a ways off the coasts of British Columbia and the State of Washington. A researcher—I think he was from NOAA—came up with a nickname for that pool of warm water: The Blob."

"The Blob? What did he do? Give it a cutesy name so no one would think it could have contributed to global warming? Because, once again, it sounds like Mother Nature was responsible for a warming event that impacted temperatures globally."

"Two warming events caused by Mother Nature, sweetie—the El Niño and The Blob— running in parallel but also independent of one another in time, space, and process. More to the point, The Blob formed mid-year in 2013, and the El Niño evolved in late in 2014. The El Niño decayed in 2016. The Blob dissipated about the same time. But all of that warmer-than-normal water created by The Blob didn't simply disappear. All the while the Blob existed, that warmer-than normal water was being distributed by ocean currents, down the coasts of California and Mexico."

"Thanks. That clarifies things. But to continue my thought, a shift that large in the East Pacific, which covers 30% of the surface of the global oceans, had to have had an impact on global surface temperatures."

"Whoa there, Anna. I'm not disagreeing with you. Let me make that perfectly clear. I agree wholeheartedly with what you said. But think about what you just said. If I had said a couple of days ago what you said just now, you would've called me a denier right there on the spot."

"Well, Dad, you're probably right. But..." I shrugged and grinned "...times and opinions change. Thanks to you. With what we discussed yesterday about how poorly climate models actually perform, and about how the temperature data for the surfaces of the oceans show they warmed naturally since 1982, I'll never worry about catastrophic human-induced global warming again."

"I was hoping you'd eventually come to say something like that. Nothing good ever comes from worrying about things we can't control."

I answered him by giving him a big hug.

With me in his arms, Dad said, "Okay! Why don't we go down and make some chili and cornbread for lunch. Later on, I'll email you some links for a news article and a scientific paper about The Blob."

# # #

And he did. The email with those links follows.

Morning, Anna. Here are the links I promised:

Scientific paper: Causes and impacts of the 2014 warm anomaly in the NE Pacific[16] *(2015) by Bond et al.*
A June 2014 newsletter[17] *from Bond, the Oregon State Climatologist*

Even NOAA talked about The Blob in one of their ocean briefings. I can't find the full report but here's a copy of the one page[18].
And USA TODAY'S article How the blob caused USA's weird weather[19] *from April 2015.*

Happy reading.

Love you

Dad

# # #

Dad returned to the topic of sea level. "I mentioned yesterday that comparisons of tide gauge-based sea level data and satellite-based sea level data can be misleading."

"Yes you did. You showed me how misleading it was when alarmists presented graphs that compared the trend of the long-term tide gauge-based sea level data with the trend of the short-term satellite-based sea level data. They hid the fact that there were multidecadal changes in the trends of the long-term tide gauge-based sea level data, and that there were 21-year periods when sea levels rose at rates that were comparable to and faster than the rate of the satellite-based data."

"Again, a great summary, Anna. But there were other ways those simple trend comparisons can be misleading."

"Oh yeah."

"First way, tide gauges measure sea levels along shorelines, but satellites measure sea levels of the oceans globally, not only along the shorelines but out thousands of miles from shore, too. To add to the complexities, there are major differences in rates of sea level change—rising in some locations, falling in others—throughout the oceans and along the shorelines. Unless the satellite-based measurements are subdivided and furnished for the same coordinates as the tide gauges, comparisons like the one in that graph are not accounting for those differences, Anna. Regardless, doom mongers create and publish graphs that compare them regularly."

"Sure. And the reason they do it is to show the difference in the long-term and short-term trends, all the while hiding the fact that sea levels rose faster during the early-to-mid 20th Century than they have during the satellite era."

"Agreed, Anna. The second point to consider is that satellites are calibrated against tide gauges, so along the shorelines the concurrent trends of the raw satellite data *should be* similar to those of the tide gauges.

"A third point: another difference between the

two datasets is that the satellite data receive additional adjustments to account for factors such as post-glacial rebound."

I noted, "Post-glacial rebound results from the melting of ice sheets and glaciers many millennia ago. Without the weight of the glaciers and ice sheets, land surfaces are springing back up, rebounding, in parts of the world where the ice existed during the ice age."

"But you're overlooking an important aspect. And it's the reason why scientists add to the rates of satellite-observed sea level rise to account for it. While the land surfaces are rising with the loss of the ice mass, the ocean floors are sinking. So scientists adjust the sea level data upwards with time to account for those factors, and those adjustments create higher rates of global sea level rise."

"But that sounds logical, Dad."

"Adjusting satellite-derived data for post-glacial rebound and other factors may be important to scientists. But are they relevant to the problems many shoreline communities will encounter with rising sea levels? No. Nobody, other than specialized scientists, cares about the sinking ocean floor out in the center of the oceans. Sea level is a localized problem facing shoreline communities."

"No argument from me on that point."

He continued, "There are even more considerations about sea levels being local, not global, problems: the ground levels are sinking in some shoreline locations due to the removal of ground water and oil. Plate tectonics play roles in other spots, with ground levels rising in some locations and sinking in others."

I asked, "Do you have a model-data comparison for sea levels?"

"Nope. Sorry, Anna. The model outputs of global mean sea level are not available from the KNMI Climate Explorer. At least, I've never found them."

"Too bad. I wanted to see how poorly they simulated sea levels. Are they as bad with sea levels as they are with global surface temperatures, global precipitation and global sea ice?"

"Sorry again. Without the model outputs, I'd only be speculating."

THE PATHETIC POLITICS BEHIND THE GLOBAL WARMING MOVEMENT

After lunch, we resumed our discussions of global warming and climate change, but this time it was the politics behind the movement.

Dad began, "Now, I'll bet you're wondering, Anna, how the science behind the global warming scare can be so bad."

"The thought has occurred to me over the past two days."

"Why don't we start our discussions of the politics with a look at a portion of Prime Minister Margaret Thatcher's memoire *Statecraft*, to read her thoughts on this global warming nonsense? Unlike every other politician I can think of, she had the courage to change her public stance on global warming. But by the time she wrote *Statecraft*, she was long out of office."

He walked briskly to a well-stocked bookshelf,

and plucked a copy of Margaret Thatcher's book from the second shelf. He flipped through the pages as he retook his seat. "Here we are, under the heading of *Hot Air and Global Warming.* Yup, lots of hot air from that lot of blowhards."

In college, at the suggestion of a roommate, I had read one of Margaret Thatcher's earlier memoires *The Downing Street Years,* published in 1993. So Thatcher's words a decade later in *Statecraft* were eye-opening. That's because they were far different from my memories of her as a climate hawk during her years as Britain's Prime Minister.

"She wrote in 2002 in *Statecraft.*" My Dad then read aloud:

> *The doomsters' favorite subject today is climate change. This has a number of attractions for them. First, the science is extremely obscure so they cannot easily be proved wrong. Second, we all have ideas about the weather: traditionally, the English on first acquaintance talk of little else. Third, since clearly no plan to alter climate could be considered on anything but a global scale, it provides a marvellous excuse for worldwide, supra-national socialism.*
>
> *All this suggests a degree of calculation. Yet perhaps that is to miss half the point. Rather, as it was said of Hamlet that there was method in his madness, so one feels that in the case of some of the gloomier alarmists there is a large amount of madness in their method. Indeed, the lack of any sense of proportion is what*

characterises many pronouncements on the matter by otherwise sensible people. Thus President Clinton on a visit to China, which poses a serious strategic challenge to the US, confided to his host, President Jiang Zemin, that his greatest concern was the prospect that 'your people may get rich like our people, and instead of riding bicycles, they will drive automobiles, and the increase in greenhouse gases will make the planet more dangerous for all.'

It would, though, be difficult to beat for apocalyptic hyperbole former Vice President Gore. Mr Gore believes: 'The cleavage in the modern world between mind and body, man and nature, has created a new kind of addiction: I believe that our civilisation is, in effect, addicted to the consumption of the earth itself.' And he warns: 'Unless we find a way to dramatically change our civilisation and our way of thinking about the relationship between humankind and the earth, our children will inherit a wasteland.'

But why pick on the Americans? Britain's then Foreign Secretary, Robin Cook, has observed: 'There is no greater national duty than the defence of our shoreline. But the most immediate threat to it today is the encroaching sea.' Britain has found, it seems, a worthy successor to King Canute.

The fact that seasoned politicians can say such ridiculous things – and get away with it – illustrates the degree to which the new dogma

> *about climate change has swept through the left-of-centre governing classes. ...”*

“Reading between the lines, Dad, I have to admit the quote from President Bill Clinton caught me off guard. He seems to be expressing his fear back then that China would become the world's leading economic superpower, taking the place of the United States.”

Dad said, “Clinton visited China at the invitation of Jiang Zemin in the summer of 1998. Nearly twenty years later, there are reports that China will surpass the United States as the world's leading economic superpower soon, if it hasn't already.”

I asked, “Please read the sentence that ends in socialism again. It was toward the beginning.”

“*...Third, since clearly no plan to alter climate could be considered on anything but a global scale, it* [climate change] *provides a marvellous excuse for worldwide, supra-national socialism...*”

“‘Worldwide, supra-national socialism’? That would fly like a lead balloon here in the States.”

“It's playtime, Anna. Imagine teaching our kids a new Pledge of Allegiance in grade school.” He put on his Walter Cronkite voice as he said, “I pledge allegiance to the flag of the Worldwide United Nations of Earth...”

I added, giggling as I did, “...and to the socialist state for which it stands, one planet under the unelected politicians of the United Nations...”

“...blah, blah, blah, oy vey.”

We both laughed uncontrolled for a minute or so.

“As I said, like a lead balloon.”

"Some people love to be *led* mindlessly throughout their lives like balloons on strings."

We sat quietly for a few moments, until I broke the silence. "The political aspects behind the global warming scare: there have to be lots of reasons, probably as many reasons as there are countries on Earth."

"The Iron Lady gives us her insights later under that heading of *Hot Air and Global Warming*:

> *By the end of my time as Prime Minister I was also becoming seriously concerned about the anti-capitalist arguments which the campaigners against global warming were deploying. So in a speech to scientists in 1990 I observed:*
>
> > *'Whatever international action we agree upon to deal with environmental problems, we must enable our economies to grow and develop, because without growth you cannot generate the wealth required to pay for the protection of the environment. We can rely on industry to show the inventiveness which is crucial to finding solutions to our environmental problems.'*

"According to the footnotes, that speech was to the Royal Society in London on 22 March 1990."

He continued, "This is the beginning of the next paragraph after her quote from the speech to the Royal Society:"

> *Since then, there have been two further developments. First, the anti-capitalism which always lay under the surface of environmentalism has become much more explicit, reflected most recently in an ugly streak of anti-Americanism. This came into full view when President Bush announced in March 2001 that the US would not sign the Kyoto protocol on climate change...*

"Hold on, Dad. Bill Clinton and Al Gore were President and Vice President when the Kyoto Protocol was signed in 1997. It's kind of odd that global politicians got nasty with George W. Bush when he only *confirmed* that the U.S. would not sign it."

"Such is global politics." Dad shrugged. "Back to what Thatcher wrote. She then continued that paragraph with an example of an anti-American quote from the Environmental Minister of France, and she mentions that the Environmental Minister of the European Union made unspecified threats against US businesses. But let's proceed to the paragraph after that. Thatcher wrote:"

> *Actually, President Bush was quite right to reject the Kyoto protocol. His predecessor* [Bill Clinton] *had supported it for international effect, while knowing its provisions made it a dead letter at home: the US Senate had voted unanimously on the matter. The protocol would have placed all the burden for reduction of* CO_2 *on developed countries, while leaving the developing countries — including India and China — to keep producing it at a rapidly*

> *growing rate. America's target for cuts was totally unrealistic — a 7 per cent reduction of overall emissions of greenhouse gases below 1990 levels between 2008 and 2012. And all this is before anyone considers the scientific arguments about why and to what extent global warming is occurring. Kyoto was an* anti-growth, anti-capitalist, anti-American *project which no American leader alert to his country's national interests could have supported.*

"'Anti-growth, anti-capitalist, anti-American'", I repeated.

"One more time. The international global warming movement is..."

"'Anti-growth, anti-capitalist, anti-American', according to the illustrious stateswoman Margaret Thatcher."

# # #

Dear reader, I won't bore you with all of our additional conversations about politics, just the highlights.

# # #

I said, "Ms. Thatcher was quite prophetic with her thoughts on China's CO_2 emissions. They rose above United States emissions in 2007."

I then changed topics, while still discussing the Iron Lady's writings, "You read the first thing that

Mrs. Thatcher said had changed, but she said there were two developments."

"Correct again, Anna." Dad once more read from Thatcher's *Statecraft*:

> *The other change which has occurred in the debate about climate change since I was in Downing Street is that the science has moved on. And, as is usually the way with scientific advance, the picture looks more rather than less complex.*

"End quote, but I need to clarify something for you, Anna. As the Iron Lady had written, the science behind global warming had, in fact, moved on, but it had not advanced. It had degenerated, and it continues in that degenerated state today."

"Now, with what we've discussed, I understand completely."

"The climate science community now claims they know, in no uncertain terms, the reason for global warming. And, according to them, it is far and away the emissions of anthropogenic greenhouse gases, not Mother Nature. But their only tools, climate models, contradict them. That is, as we've seen, comparisons of data and the model outputs show that global warming can occur without being forced by the greenhouse gases that drive the models. Those models clearly show to anyone who bothers to examine them that the foundations for climate science, the models, are nothing but multi-billion dollar trash."

"Trash is too nice a word for them, Dad."

He continued without breaking stride, "All the while, the climate scientists…a.k.a. computer programmers…claim to know all."

"Like the 'Great and Powerful Oz'."

"*The Wizard of Oz.* Nice touch, Anna," he said, chuckling.

Dad continued, "Climate science may have been respectable many decades ago, but it has been forever corrupted by politicians."

"Assuming climate science was better before politicians took control, the degeneration of climate science at the hands of politicians is a shame."

"Indeed, Anna. Now it's full of misdirection, misconceptions, misinformation…"

"Misdiagnoses, mistakes, misattributions, miscalculations," I interjected.

He smiled broadly. "Mischaracterizations, misinterpretations, missed guesses, miscomprehensions."

"My turn: mistakes, mistruths, misunderstandings. Well, they've obviously *missed* a lot."

"Right you are," Dad agreed. "They've missed a whole lot. Okay, here's my favorite bit of misinformation and misdirection. Climate scientists point to the models and say, without greenhouse gases, natural variability could not have caused the warming in the latter part of the 20th Century."

"I think we've already discussed that."

"Yes, but a picture's worth… Hold on for a moment while I search for an illustration from the IPCC's 5th Assessment Report."

I've included the IPCC's figure FAQ10.1, Figure 1

from the 5th Assessment Report of the Intergovernmental Panel on Climate Change (IPCC) below as my Exhibit 18.

Exhibit 18

FAQ 10.1, Figure 1 from IPCC 5th Assessment Report

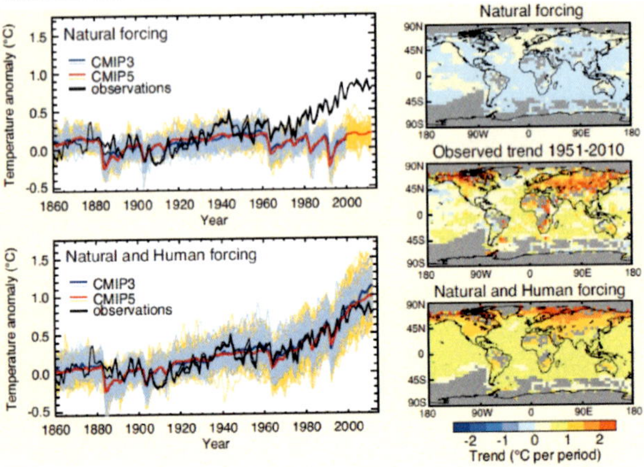

Dad said after handing me his tablet computer, “The text that accompanies that illustration rates an 11 on the BS meter scale of 10. The accompanying text from the IPCC’s 5th Assessment Report reads:”

> *FAQ 10.1, Figure 1 illustrates part of a fingerprint assessment of global temperature change at the surface during the late 20th century. The observed change in the latter half of the 20th century, shown by the black time series in the left panels, is larger than expected from just internal variability. Simulations*

> *driven only by natural forcings (yellow and blue lines in the upper left panel) fail to reproduce late 20th century global warming at the surface with a spatial pattern of change (upper right) completely different from the observed pattern of change (middle right). Simulations including both natural and human-caused forcings provide a much better representation of the time rate of change (lower left) and spatial pattern (lower right) of observed surface temperature change.*
>
> *Both panels on the left show that computer models reproduce the naturally forced surface cooling observed for a year or two after major volcanic eruptions, such as occurred in 1982 and 1991. Natural forcing simulations capture the short-lived temperature changes following eruptions, but only the natural + human caused forcing simulations simulate the longer-lived warming trend.*

Let me repeat the last part," I said, "'*...but only the natural + human caused forcing simulations simulate the longer-lived warming trend.*'"

"In other words, in the virtual worlds of the climate models, they could only reproduce the warming in the latter part of the 20th Century when they added as inputs the numerical representations of anthropogenic greenhouse gases."

"What a childish way to attribute global warming to manmade greenhouse gases. All they've show is that they can program computers to provide desired results."

"And the caption for their FAQ 10.1, Figure 1 reads":

> ***FAQ 10.1, Figure 1*** **|** *(Left) Time series of global and annual-averaged surface temperature change from 1860 to 2010. The top left panel shows results from two ensemble [sic] of climate models driven with just natural forcings, shown as thin blue and yellow lines; ensemble average temperature changes are thick blue and red lines. Three different observed estimates are shown as black lines. The lower left panel shows simulations by the same models, but driven with both natural forcing and human-induced changes in greenhouse gases and aerosols. (Right) Spatial patterns of local surface temperature trends from 1951 to 2010. The upper panel shows the pattern of trends from a large ensemble of Coupled Model Intercomparison Project Phase 5 (CMIP5) simulations driven with just natural forcings. The bottom panel shows trends from a corresponding ensemble of simulations driven with natural + human forcings. The middle panel shows the pattern of observed trends from the Hadley Centre/ Climatic Research Unit gridded surface temperature data set 4 (HadCRUT4) during this period.*

"That logic is so childish, Dad."

"I couldn't agree with you more, sweetie. But the computer model runs in that illustration and the childish logic behind them serve as their evidence, their 'fingerprints', that prove beyond a shadow of a doubt that anthropogenic greenhouse gases are responsible for global warming since the 1950s."

"That's the best they can do?" I asked, shaking my head in disbelief. "What a farce."

"What the IPCC failed to disclose is, the climate models used in those studies do not—actually cannot—simulate the naturally occurring coupled ocean-atmosphere processes that can cause global surface warming over multidecadal periods like strong El Niño events and the Atlantic Multidecadal Oscillation."

"But I've learned much over the past few days, Dad. From this moment on, I will be skeptical of all of the global warming and climate change farce."

"May the farce be with you."

"Oh, Dad." Sometimes his dad jokes...enough said.

"So now it's time to delve into all of the sordid details behind the Intergovernmental Panel on Climate Change, the IPCC."

I let a sly grin form on my face as I replied, "We've been mentioning them since we started these talks. They're portrayed in the news as an international scientific body."

"And the mainstream media is wrong again. The IPCC is a political entity, not a scientific one. The IPCC was founded by the United Nations to support political agendas, nothing more, nothing less" He tinkered with his computer, this time appearing to be surfing online. Moments later, he said, "I'll read to you from the IPCC's website."

"If you must."

"I must, I must. This is from the opening paragraph of the IPCC History[20] webpage:

> *The initial task for the IPCC as outlined in UN General Assembly Resolution 43/53 of 6 December 1988 was to prepare a comprehensive review and recommendations with respect to the state of knowledge of the science of climate change; the social and economic impact of climate change, and possible response strategies and elements for inclusion in a possible future international convention on climate.*

"As I said, Anna, the IPCC was founded to write reports. Granted, the full reports are very detailed reports, so burdensome that very few people bother to read them in their entirety. But those reports were prepared for the sole purpose of supporting political agendas." Looking up from the display, he asked, "Have you read the reports?"

I replied in all honesty, "Yes I have. I've read the last two Summaries for Policymakers."

Dad chuckled with an unnerving grin then stared holes in my eyes. "Are you aware that the language of each IPCC Summary for Policymakers is agreed to by *politicians* during week-long strategy meetings called plenary sessions?"

"No," I answered.

"A draft of the Summary for Policymakers is prepared for the politicians by a few scientists. Then the politicians—all of the politicians attending that particular conference—rewrite the entire Summary for Policymakers, after debates, and more debates, and even more debates, about the language to be included in *each* sentence and whether or not a sentence is to be included in the summary. To add icing to my distrust, those week-long political debates about the Summary for Policymakers are closed to the public. Again, Anna, global politics at its worst."

"Let me summarize. Policymakers from around the globe debate and rewrite every sentence in the IPCC Summary for Policymakers, including some but excluding others, with the intent of assuring scientific support for their own agendas."

"Correct. Policymakers write the Summary for Policymakers. In other words, they're writing nothing more than memos to themselves."

"So the only real reason for the reports is to convince the public there's a problem that needs to be fixed and paid for, even though, in reality, there might be no problem at all."

"Precisely. Now can you understand my total distrust of the IPCC and climate science in general?"

"I have to admit all of it stinks, especially after you've shown me how bad the science really is."

"You can verify online how the summaries for policymakers are prepared."

# # #

And I did at a later date. The following is a small portion of the overview from the IPCC webpage titled Summary Description of the IPCC Process[21], with the appropriate portion underlined:

In parallel, the Summary for Policymakers and the Technical Summary are prepared by authors selected from chapter writing teams. The Summary for Policymakers and the Technical Summary undergo a formal Expert and Government Review and are revised based on the review comments. Again, 2-3 Review Editors oversee the review process for the Technical Summary. The Summary for Policymakers is then distributed to the Governments for final comment. The Summary for Policymakers undergoes a line-by-line approval process by Governments during a multi-day meeting. The Governments must reach consensus before text is approved. After the Working Group Session approves the Summary for Policymakers, the IPCC Plenary then accepts the entire underlying Report (including the individual chapters and Technical Summary).

# # #

I glanced at Dad who appeared to be examining files on his computer. A question struck me, so I asked, "But what of the full scientific assessments? Those are thousands of pages long."

His playful fatherly smirk appeared. "Does the additional verbiage make them correct? Let me answer my own question with a resounding and protracted *no.* The climate scientists who prepare them are nothing more than funding-hungry toadies to the politicians. No one preparing those nonsensical boat anchors would include scientific studies that contradicted the politically financed groupthink. In fact, scientists would lose their funding if their studies disagreed with the agendas of the politicians who supply the moneys."

"Sounds like you're saying there's no scientific integrity."

"In climate science? Surely not. Think of the poor scientists who do not support the political agendas, those rare few who actually have independent thoughts and do not adhere to the groupthink. They are turned into pariahs by their politician-subservient peers. The free-thinkers are labeled as deniers—to link them to holocaust deniers—and the politicians cut off their funding. But the outspoken pseudo-scientists slash funding leeches who *do* support the politicians can become mainstream-media darlings…a climate-related form of the regressive human parasites who make up the species called reality-TV celebrities."

"You've been using the word groupthink in place of consensus, Dad."

"Yup, every time I hear the word consensus, I think groupthink."

"I'm beginning to understand why."

"A consensus is a political goal, not a scientific one, Anna. Science is a search for correct answers, not agreement on those answers."

"Okay, for climate science, using groupthink in place of consensus makes a lot of sense. Groupthink stifles the search for correct answers."

"Now here's a real problem. It's really, really

unlikely there will be a flip-flop on the groupthinks about human-induced global warming and climate change, even though there are an ever-growing number of failed prognostications and even more failings with sustainable sources of energy. And even if the general public really understood how horrendous the science actually was, there are too many political and business reputations at stake and far too many billions of dollars involved for politicians to backpedal, at least if they want to remain in office."

Dad began drawing on his unlit pipe, then said, "Recall something from our earlier reading from the IPCC website, the stuff about their founding the IPCC in 1988. There was mention of a 'possible future international convention on climate'. That international treaty must have already been in the oven because it was signed four years later in 1992. It was called the UNFCCC or the United Nations Framework Convention on Climate Change."

"I've heard of the UNFCCC, Dad. It was the foundation for the Kyoto Protocol and for the more-recent Paris Agreement."

"Right again you are." He puttered around on the touchscreen monitor and then continued, "You'll find this very interesting. It's the opening paragraph from the United Nations' UNFCCC background webpage[22]." Handing me the tablet computer, he asked with a mischievous smile, "Please read it aloud." So I did.

> *In 1992, countries joined an international treaty, the United Nations Framework*

> *Convention on Climate Change, as a framework for international cooperation to combat climate change by limiting average global temperature increases and the resulting climate change, and coping with impacts that were, by then, inevitable.*

"What's odd about that statement?" Dad asked.

"Nothing. It appears straightforward."

"So it does, unless you know the history of the IPCC. The IPCC released their First Assessment Report in 1991, a year before the UNFCCC was signed. But in the First Assessment Report by the IPCC[23], the climate science community stated very clearly that they could not differentiate between human-induced and natural global warming. And they further claimed it would take a decade or more before they might be able to detect any human-induced contribution. Yet, a year after the scientists said they couldn't attribute the warming to human activities, the politicians were working at, '*limiting average global temperature increases and the resulting climate change, and coping with impacts that were, by then, inevitable.*'" He reached out his hand and asked, "My computer, please."

His fingers danced again on the keyboard. Dad looked overjoyed to have me as an audience. "Alas," he said. "Should only take a few moments to reload. Let me do a quick word search for the word *judgement*." He tapped and waited a few times, said, "Now to confirm what I just said. This is from the Summary for Policymakers of the IPCC's First Assessment Report." He then read aloud:

"In boldface, '***Our judgement is that:***

"First bullet point: *'Global-mean surface air temperature has increased by 0 3°C to 0 6°C over the last 100 years, with the five global-average warmest years being in the 1980s. Over the same period global sea level has increased by 10-20cm. These increases have not been smooth with time, nor uniform over the globe.'*

"Twenty centimeters is roughly 7 inches, sweetie. I believe people throughout most of the 1900s have managed to outrun that 7 inches of rising seas. Maybe we're slower now?"

"Sarcastic point taken, Dad."

"Second bullet point: *'The size of this warming is broadly consistent with predictions of climate models, but it is also of the same magnitude as natural climate variability. Thus unequivocal detection of the enhanced greenhouse effect from observations is not likely for a decade or more[sic]'*

"End quote. Thus my statement that the IPCC could not differentiate between natural and human-induced global warming one year before the UNFCCC treaty was signed."

I added, "Even though, based on the science at that time, it might have been a fight solely against Mother Nature. Odd, Dad, very odd. The more you present, the more bizarre the global warming scare sounds."

"And with their statement that it could take a decade or more to be able to unequivocally detect the advanced greenhouse effect from data, the scientists were prying open the politicians' pocketbooks for funding without firm time limits."

"Yeah, I caught that."

"It gets even more bizarre, my beautiful Anna." I must have looked bewildered because he added, "Yes, even more bizarre."

"You've got to be kidding me."

"Ready for this? In 1995, four years after the first report, three years after the international climate-fighting UNFCCC treaty was signed, the politicians' report-writing entity, the IPCC, still could not differentiate between man-made and natural warming."

"No," I said, dragging out that simple word.

"And the climate scientists had stated that fact in the first draft of the Second IPCC assessment report, in the summary for policymakers and in the main report. Imagine that. The politicians had to be twitching and shaking in their foolishness. They'd had an international treaty in place for 3 years, but their report-writing scientists still could find no evidence to support it. In other words, the IPCC had failed in their prime directive, which was to write reports that supported the politicians' international treaties. By that time, the politicians had to have even drafted a treaty called the Kyoto Protocol, to be signed a few years later, in 1997. It was a treaty to limit emissions of anthropogenic greenhouse gases." He snickered. "Yet the first draft of the IPCC's second report in 1995 said the scientific community could not find the required evidence to support it. So what do you suppose happened next?"

"I know the Kyoto Protocol[24] was adopted; that is, it was signed by 150 countries. That number stands out in my mind. And if memory serves, I'll agree with you that the year was 1997. So the politicians must have ignored the IPCC's second assessment report from two years earlier."

"Please, Anna, we're speaking about politicians...you know, prevaricating, insecure, egocentric people with Messiah complexes."

"Oof, that was harsh."

"But realistic. No, they didn't ignore that significant roadblock to their treaties. The politicians simply exerted enough pressure on the scientists so that the scientists changed the language of the offensive portions of the Summary for Policymakers, and of the full report, so that the reports supported the political agendas."

My mouth must've dropped open. I was dumbfounded. "Really, truly?"

"Really, truly, Anna. Don't you recall all the hubbub back then with some of the scientists complaining about the changes made to that report? They were furious and told the story in newspaper and magazine op-eds and interviews. Other scientists were downplaying the changes. Thus, the first lines were drawn at that time between the skeptical scientists and the funding-hungry puppets of the politicians."

"Sorry, Dad, I was too young back then. Even if I wasn't, like most people, I wouldn't have been paying attention to global warming and climate change back the 90s."

"Sorry. You're right, sweetie. For the rest of the world, the primary reason that global warming wasn't a hot item back then: those topics hadn't yet become the darlings of the mainstream media."

"Now, though, it seems a day can't go by without a story about some weather-related catastrophe somewhere in the world or some doomsday forecast based on some climate model study."

"Chicken Little would be proud." Dad paused before continuing, "Sadly, Anna, for you to be able to confirm what I recall about the first draft of the IPCC's second assessment report, you'd have to search newspaper archives to find the stories." Dad once again steepled his fingers, then tilted his head. His reflections were short-lived. "On second thought, I'll send you links to a very detailed three-part article about that monumental change to the language of the IPCC's Second Report. It is very in-depth, and the happenings at that conference are so involved and have to do with so many people from so many countries that it would be impractical for us to dig into it now. I'll email you hyperlinks, and you can study them whenever you want."

# # #

Note to my readers: Dad did in fact email the hyperlinks. (Hopefully, they'll work for you.) See the 3-part, very detailed article by Bernie Lewin about the 1995 IPCC conference in Madrid:

Part 1- Madrid 1995: Was this the Tipping Point in the Corruption of Climate Science?[25]

Part 2 – Madrid 1995 and The Quest for the Mirror in the Sky[26]

Part 3 – Madrid 1995: The Last Day of Climate Science[27]

Dad also included a link to a recently published book by Bernie Lewin, Searching for the Catastrophe Signal: The Origins of The Intergovernmental Panel on Climate Change[28].

# # #

My father said, “Bottom line: the politicians dictated the science.”

I huffed in utter disgust then said, including a small portion of one of the quotes from Margaret Thatcher, “Well, it sounds like, without the changes to the Second Assessment Report, the politicians were afraid their ‘anti-growth, anti-capitalist, anti-American’ policies would disintegrate, and most of the scientists were afraid that the IPCC would die and all the funding for carbon dioxide-based global warming research would disappear.”

“That sounds like a reasonable summary.”

“Basically, climate science was preordained by the politicians.”

“No doubt in my mind,” Dad replied. “And the politicians continued to throw money at the climate scientists/toadies so they’ve had a continuous source of funds for their worthless science…”

“Funds that they never would have had if it wasn’t for the global warming hysteria created by the politicians.”

“And the timing fits perfectly with the 1991 collapse of the Soviet Union, so that the mainstream media had a new evil to profit from."

I had nothing to add at that moment.

# # #

"Remember how poorly the latest batch of climate models simulated the global surface warming that occurred from the 1910s to the 1940s?"

"Yup, the warming was more than 3 times faster than hindcast by the models."

"That's always been a problem for the models." He opened a .pdf file on his computer. "Let me see if I can remember the figure number of the model-data comparison from the IPCC's First Assessment Report. Got it. Here we are: Figure 8.1."

The IPCC's Figure 8.1 from their First Assessment Report is included below as my Exhibit 19.

I belly laughed and it took me about 10 seconds to regain control. "Those early models were horrendous. The models came nowhere close to being able to simulate the warming from about 1910 to the 1940. Didn't they say something about the global warming being consistent with the models?"

Exhibit 19 - IPCC FAR Figure 8.1

246 *Detection of the Greenhouse Effect in the Observations 8*

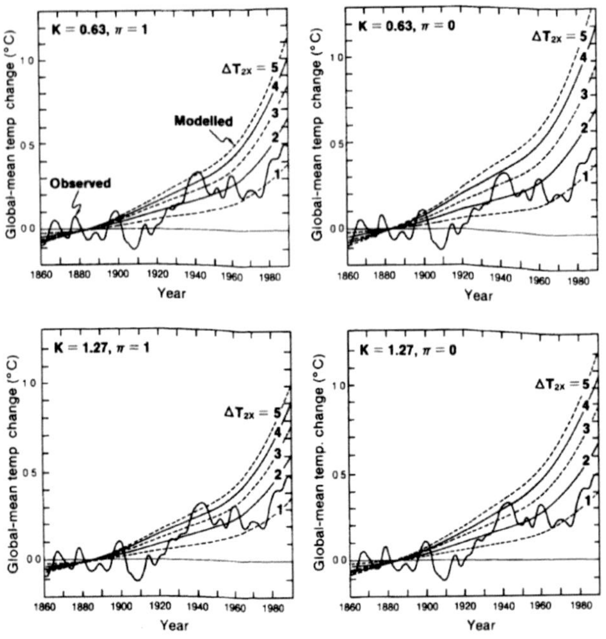

Figure 8.1: Observed global-mean temperature changes (1861 1989) compared with predicted values The observed changes are as given in Section 7 smoothed to show the decadal and longer time scale trends more clearly Predictions are based on observed concentration changes and concentration/forcing relationships given in Section 2 and have been calculated using the upwelling diffusion climate model of Wigley and Raper (1987) To provide a common reference level modelled and observed data have been adjusted to have zero mean over 1861 1900 To illustrate the sensitivity to model parameters model results are shown for ΔT_{2x} = 1, 2, 3 4 and 5°C (all panels) and for four K, π combinations The top left panel uses the values recommended in Section 6 (K = 0.63cm²sec⁻¹, π = 1) Since sensitivity to K is relatively small and sensitivity to π is small for small ΔT_{2x}, the best fit ΔT_{2x} depends little on the choice of K and π

"Yup, they said in the Summary for Policymakers:

- *Global-mean surface air temperature has increased by 0 3°C to 0 6°C over the last 100 years, with the five global-average warmest years being in the 1980s. Over the same period global sea level has increased by 10-20cm. These increases have not been smooth with time, nor uniform over the globe.*

- *The size of this warming is broadly consistent with predictions of climate models, but it is also of the same magnitude as natural climate variability...*

Looking at their model-data comparison again, I interrupted, "Broadly consistent? The simulations come nowhere close to simulating the variability. It looks like they've tried to overlay exponential curves on a sine wave curve and claim they matched."

"Trying to match exponential curves to a sine wave is a good analogy. Thanks for that, Anna."

"Those models were pathetic, Dad."

"Agreed. And they covered their asses, or tried to, with 'but it is also of the same magnitude as natural climate variability'. Let me continue the second bullet point:

> "'*...Thus the observed increase could be largely due to this natural variability, alternatively this variability and other human factors could have offset a still larger human-induced greenhouse warming The unequivocal detection of the enhanced greenhouse effect from observations is not likely for a decade or more'*

"But the supposed 'unequivocal detection of the

enhanced greenhouse effect from observations' came a few years later in the Second Assessment Report," I recalled aloud. "Do you have a similar graph from that Assessment Report?"

"Of course. Give me a few seconds." He opened another .pdf file stored on this computer. "This time I think it was their Figure 8.4." After a few moments, he continued, "Yup, here it is."

# # #

That illustration from the IPCC's Second Assessment Report had three cells that were stacked, and when I tried to upload it to this book, the text was clipped off the bottom of the caption. So I've reproduced all three cells individually below in my Exhibits 20a, 20b, and 20c, including the full caption in all three.

Exhibit 20a - IPCC SAR Figure 8.4 Cell A

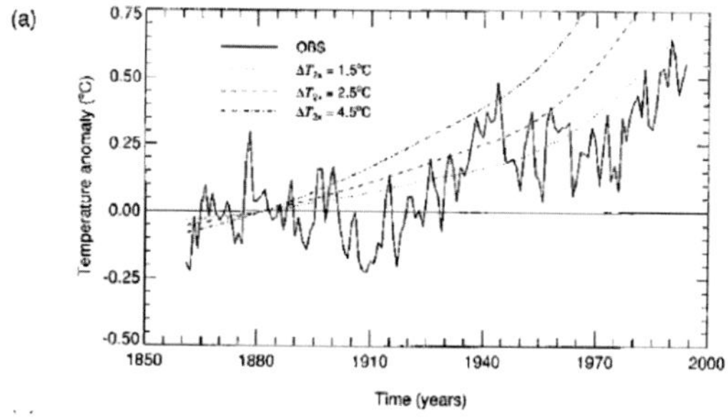

Figure 8.4: Observed changes in global mean temperature over 1861 to 1994 compared with those simulated using an upwelling diffusion-energy balance climate model. The model was run first with forcing due to greenhouse gases alone (a), then with greenhouse gases and aerosols (b), and finally with greenhouse gases, aerosols, and an estimate of solar irradiance changes (c). The radiative forcings were the best-guess values recommended in this report. Simulations were carried out with climate sensitivities ($\Delta T_{2\times}$) of 1.5, 2.5 and 4.5°C for the equilibrium CO_2-doubling temperature change.

Exhibit 20b - IPCC SAR Figure 8.4 Cell B

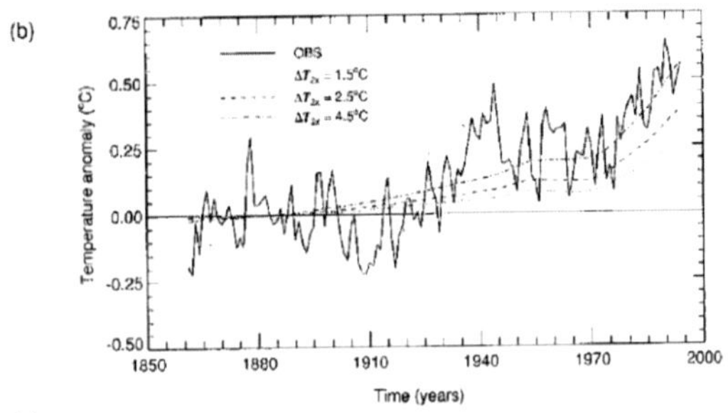

Figure 8.4: Observed changes in global mean temperature over 1861 to 1994 compared with those simulated using an upwelling diffusion-energy balance climate model. The model was run first with forcing due to greenhouse gases alone (a), then with greenhouse gases and aerosols (b), and finally with greenhouse gases, aerosols, and an estimate of solar irradiance changes (c). The radiative forcings were the best-guess values recommended in this report. Simulations were carried out with climate sensitivities ($\Delta T_{2\times}$) of 1.5, 2.5 and 4.5°C for the equilibrium CO_2-doubling temperature change.

Exhibit 20c - IPCC SAR Figure 8.4 Cell C

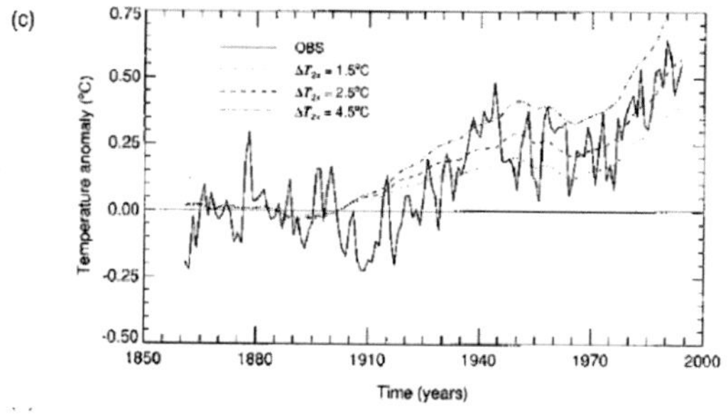

Figure 8.4: Observed changes in global mean temperature over 1861 to 1994 compared with those simulated using an upwelling diffusion-energy balance climate model. The model was run first with forcing due to greenhouse gases alone (a), then with greenhouse gases and aerosols (b), and finally with greenhouse gases, aerosols, and an estimate of solar irradiance changes (c). The radiative forcings were the best-guess values recommended in this report. Simulations were carried out with climate sensitivities ($\Delta T_{2\times}$) of 1.5, 2.5 and 4.5°C for the equilibrium CO_2-doubling temperature change.

# # #

Dad said, "The intent of the illustration is obvious. The model simulations of surface temperatures are terrible if they use only greenhouse gases to drive the models, as shown in Cell a. And as shown in Cell b, the models improve a little in the latter part of the 20th Century, if they suppress the mid-20th Century warming in the models with numerical representations of aerosols from burning fossil fuels."

I interjected about Cell c, "And they get the best fit from about 1950 to 1994 if they add variations on solar irradiance to the greenhouse gases and aerosols, but it only gets better during that timeframe. And no matter how high they tweak the climate sensitivity to a doubling of CO_2, the models can't come close to simulating the warming from the 1910s to the 1940s."

"But when they tweak it to better match the early warming from the 1910s to the 1940s," Dad observed, "the models warm too fast for the second warming from the 1970s to the 1990s, and the models climb off the chart—way, way ahead of the actual warming shown by the data."

It was my turn to talk about the trash outputs of the climate models. "And that problem with the models being unable to explain the warming in the early 20th Century still exists today. Yet it was that early batch of models they were referring to when the climate scientists claimed they could now detect man's influence on climate. Just because they got the models to better align with the data in the second half of the 20th Century? Yikes! But hey, you guys, what about the warming in the first half of the 20th Century?"

"Always keep in mind, sweetie, this is politics, not science. Are you expecting politics to make sense?"

"I hope it does, Dad. Well, at least on things other than climate science I hope it does. I used to believe climate science was respectable. But my beliefs have forever been changed."

"Glad to hear that, Anna. Did I say something yesterday about the IPCC using the term 'tuning knobs' when discussing climate models in the Second Assessment Report?"

"Yes you did."

"Give me a second or two, and I'll find that paragraph. Bingo, it's on the preceding page, under the heading of *8.4.1.3 Empirical estimation of the sensitivity of climate to forcing* and I quote:

> *A related univariate approach attempts to use observational data to deduce* $\boldsymbol{\Delta}T_{2x}$, *the so-called climate sensitivity parameter - a measure of how sensitive the climate system is to external forcing.' This is done by running a series of EBM simulations with different sensitivity values and choosing the value that gives a best fit with observed global mean changes. In such experiments there are a number of* '**tuning knobs**' *other than the sensitivity, both for the input historical forcings and for the EBM parameters. A few papers have considered the joint optimisation of* $\boldsymbol{\Delta}T_{2x}$. *and one Or more forcing factors, such as solar variability and/or sulphate aerosol forcing (e.g., Kelly and Wigley, 1992; Schlesinger et al., 1992; Schlesinger and Ramankutty, 1992).*

"End quote."

# # #

Note to readers, each time my Dad reached the notation "ΔT_{2x}", he read it as, "the change in global temperature in response to a doubling of atmospheric carbon dioxide levels, also known as climate sensitivity".

# # #

"Even with all of those tuning knobs, they still can't simulate the warming that took place from the 1910s to the 1940s," I added.

"And that warming was comparable—not equal to—but comparable to the warming we've seen since the mid-1970s."

"And, Dad, you've convinced me that the satellite-era sea surface temperature records indicate the warming since the early 1980s could be caused naturally by sunlight-fueled El Niño events and the Atlantic Multidecadal Oscillation."

"Bingo once again."

There was a low-key rumbling and clattering from the other side of the house, the sounds of the garage door opening.

"Your Mom is home a day early," my Dad said.

"That's great. Sadly, I think the lessons about global warming have ended then."

"Any time you want to continue, just let me know."

"I suppose you've also got data for floods, wildfires, droughts, heat waves, hurricanes and the like that will contradict the alarmist climate change crap we hear on the news."

"Yup, I've got that stuff for the United States.

Come on, let's go downstairs and greet your Mom."

As we made our way down the back stairs toward the kitchen, the sound of an opening door, the one leading from the garage to the kitchen, preceded my Mother calling, "We're home."

"We're?" I muttered to my dad, who shrugged, but we both knew who was with my Mom.

My Dad and I were joined in the kitchen by my Mom and her brother, my Uncle Marty. After hugs all around, the Q and A began. I've skipped most of them.

"What have you two been up to since you arrived home?" Uncle Marty asked.

"Visiting with Dad. Talking science fiction," I replied.

"Science fiction?" Marty asked. "I don't like science fiction."

"Lots of it on the tube now days," my Dad interjected then gave me a wink that the other two weren't able to see.

"Strangest things," I said to Mom.

Mom looked a little confused for a moment before she said, "I thought that show was called *Stranger Things*."

"My mistake," I said.

"Well, my cold was brought on by global warming," Uncle Marty interjected, going off on a tangent as was his norm. He sounded as though he was intentionally speaking through his nose. He followed that with a weak cough that definitely sounded forced.

For the remainder of the weekend, Dad and I did not get a chance to continue our discussions of the politics behind the global warming scare or the model-supported science that serves as its dubious foundation.

Regardless, when I returned to my apartment on Sunday evening, I had a new understanding and skepticism of the politics that drives the global warming movement and environmentalism in general. Still resounding in my mind were Margaret Thatcher's words: "*...since clearly no plan to alter climate could be considered on anything but a global scale, it provides a marvellous excuse for worldwide, supra-national socialism.*"

Maybe someday in the future I'll ask Dad to lend me his copy of Thatcher's *Statecraft* so I can ponder more of her insights into the anti-capitalism movement and other topics.

I was even more disturbed with how poorly climate models actually simulated Earth's climate, its past and present surface temperatures, its sea ice, its precipitation. I still have memories of Dad's sea level trends graph that showed, based on tide-gauge data, sea levels have risen over multidecadal periods in the past at rates that were comparable to and exceeded the rate shown during the satellite era. And, of course, there's the medical study that showed that cold environments cause twenty times more deaths than warm ones.

Dad and I continued our talks in the months that followed. Maybe I'll retell those conversations in another short story...or two...or three.

THE END

# # #

AUTHOR'S CLOSING NOTES

Thank you for reading *Dad, Why Are You A Global Warming Denier?*

I also have to thank the very dear friend who edited this short story. She made it much, much easier to read. As in the past, she has asked to remain anonymous due to the political nature of the global warming debate. Any remaining typos and grammatical errors are mine, not hers.

> **Update for print version**: Speaking of errors. Yes, I've discovered that there was no Exhibit 13. Anna must've been superstitious. Well, that's my excuse.

I have written extensively on El Niño and La Niña events at my blog Climate Observations[29]. Most of those posts were cross-posted at WattsUpWithThat[30] "the world's most-viewed site on global warming and climate change". If you'd like more information on the coupled-ocean atmosphere processes that drive El Niño and La Niña events, please refer to my free ebook ***Who Turned on the Heat? – The Unsuspected Global Warming Culprit, El Niño-Southern Oscillation*** – 550+ Pages, 23MB .pdf (**Free copy here**[31]).

And I've written much more about the so-called science behind the hypothesis of human induced global warming. See my free ebook ***On Global Warming and the Illusion of Control – Part 1***. Also in .pdf. (**Free copy here**[32]).

Thank you again.

Ciao!

REFERENCES

As noted earlier, the Kindle edition of this short story includes hyperlinks. In this print edition, the text associated with those hyperlinks has been numbered (superscripts), and the http addresses for those links are as follows.

1 - http://www.psmsl.org/
2 - http://sealevel.colorado.edu/
3 - https://cmip.llnl.gov/cmip5/
4 - https://climexp.knmi.nl/start.cgi?id=someone@somewhere
5 - https://www.esrl.noaa.gov/psd/data/gridded/data.gpcp.html
6 - https://www.ncdc.noaa.gov/oisst
7 - http://www.thelancet.com/journals/lancet/article/PIIS0140-6736(14)62114-0/abstract
8 - https://www.usatoday.com/story/weather/2015/05/20/cold-weather-deaths/27657269/
9 - http://data.myworld2015.org/
10 - http://berkeleyearth.org/
11 - http://origin.cpc.ncep.noaa.gov/products/analysis_monitoring/ensostuff/ONI_v5.php
12 - http://onlinelibrary.wiley.com/doi/10.1029/2006GL026894/epdf
13 - https://www.metoffice.gov.uk/hadobs/hadisst/
14 - https://www.esrl.noaa.gov/psd/people/gilbert.p.compo/CompoSardeshmukh2007a.pdf
15 - http://abcnews.go.com/US/TenWays/story?id=3719791&page=1
16 - http://onlinelibrary.wiley.com/doi/10.1002/2015GL063306/full

17 - http://www.climate.washington.edu/newsletter/2014Jun.pdf
18 - https://bobtisdale.files.wordpress.com/2016/01/figure-2-noaa-ocean-briefing-pg-25.png
19 - https://www.usatoday.com/story/weather/2015/04/20/blob-weird-weather/26073153/
20 - https://www.ipcc.ch/organization/organization_history.shtml
21 - https://wg1.ipcc.ch/statement/WGIsummary22122009.html
22 - http://unfccc.int/essential_background/items/6031.php
23 - https://www.ipcc.ch/ipccreports/far/wg_I/ipcc_far_wg_I_full_report.pdf
24 - http://unfccc.int/kyoto_protocol/items/2830.php
25 - https://enthusiasmscepticismscience.wordpress.com/2012/04/21/madrid-1995-was-this-the-tipping-point-in-the-corruption-of-climate-science/
26 - https://enthusiasmscepticismscience.wordpress.com/2012/07/01/madrid-1995-and-the-quest-for-the-mirror-in-the-sky/
27 - https://enthusiasmscepticismscience.wordpress.com/2012/08/07/madrid-1995-the-last-day-of-climate-science/
28 - https://www.amazon.com/Searching-Catastrophe-Signal-Origins-Intergovernmental-ebook/dp/B077N36Q3Z
29 - https://bobtisdale.wordpress.com/
30 - https://wattsupwiththat.com/
31 - https://bobtisdale.files.wordpress.com/2016/05/v2-tisdale-who-turned-on-the-heat-free-edition.pdf
32- https://bobtisdale.files.wordpress.com/2015/11/tisdale-on-global-warming-and-the-illusion-of-control-part-1.pdf

13975627R00076

Made in the USA
Lexington, KY
04 November 2018